T0403337

World Biodiesel Policies and Production

World Biodiesel Policies and Production

Edited by

Hyunsoo Joo
Ashok Kumar

CRC Press is an imprint of the
Taylor & Francis Group, an **informa** business

CRC Press
Taylor & Francis Group
6000 Broken Sound Parkway NW, Suite 300
Boca Raton, FL 33487-2742

CRC Press is an imprint of Taylor & Francis Group, an Informa business

Printed on acid-free paper

International Standard Book Number-13: 978-0-367-24444-6 (Hardback)

Library of Congress Cataloging-in-Publication Data

LoC Data here

Visit the Taylor & Francis Web site at
www.taylorandfrancis.com

and the CRC Press Web site at
www.crcpress.com

Contents

Preface

Over the last decade, biodiesel has repositioned its status in the renewable fuel market. Since the first production in the late 1980s, biodiesel had enjoyed abundant support from governments in many countries in the form of tax incentives, subsidies, and discounts in efforts to reduce the dependence on foreign oil and to diminish greenhouse gas emissions from fossil fuel. These supports lasted until 2006 when one of the biodiesel flagship countries in European Union (EU), Germany, claimed to remove the tax credits for biodiesel and to reduce the discount for biodiesel prices, which was finally abolished in 2007. It created a ripple effect throughout other EU countries including France and Italy, other large biodiesel producers in the EU. The United States (US) also ended the biodiesel tax incentive that made its price competitive with conventional diesel fuel in the market place in 2013. These measures directly impacted the biodiesel market, and resulted in reduction of biodiesel production.

Furthermore, in 2010, the EU Commission announced that the projected biofuels consumption level of 5.6% of transport fuel in 2020 has no significant impact on environmental sustainability when considering the effects of indirect land use changes (ILUC). It even stated that above the 5.6% share, ILUC emissions could rapidly increase potentially eroding the environmental sustainability. Numerous studies indicated that biofuel emissions are significant possibly exceeding those from conventional fuel if ILUC is included in the emission calculation. While the debate may put the biofuel industry in turmoil and make it hard for biofuel producers to plan for the future, it was interesting to observe that EU bioethanol producers were reported to have come out in support of accounting for ILUC, because, bioethanol stands to gain an advantage over biodiesel when land-use impacts are factored in. Along with the ILUC debate, the second-generation biodiesel has gained much attention in the last decade due to its non-food and non-ILUC advantages. Biodiesel production from nonfood biodiesel feedstock such as animal fats, waste cooking oil, jatropha and algae has been increasing recently in the EU and the US. In addition, there have been anti-dumping duties imposed by the EU on the biodiesel imports from the US, Indonesia, and Argentina that caused significant impacts on domestic and international oil production.

These drastic changes in biodiesel policies, feedstock and trades in the EU and the US in the last decade made it difficult to predict the future status of biodiesel. In this book, we review the development of biodiesel policies in the world over a decade from 2006 through 2017, and compare them chronologically and also thematically. Readers will have a better understanding of biodiesel policies all over the world, and their impacts on economy and technology. The backgrounds of drafting biodiesel policies, and the processes of revision and termination of the policies are explained and cross referenced with the bioethanol policies in different countries. Thus, a link has been discussed among different variables affecting the biodiesel policies and production.

This book consists of eight chapters, and the content of each chapter is as follows.

Chapter 1 introduces a brief summary of recent changes in biodiesel policy, production and consumption. It explains the need for an overview on the development of biodiesel policies over the last decade.

Chapter 2 presents the details of policies and their development in major countries over the last decade. Backgrounds of legislation on biodiesel and the impacts are summarized chronologically. The chapter provides the snapshots of biodiesel policies of various countries and their chronological changes to help readers understand what factors are considered in drafting new policies or revising the existing policies.

Chapter 3 focuses on current policies of the biodiesel program and examines the critical constraints and impediments in its path. It is interesting to note that India is encouraging biodiesel production from nonfood feedstocks.

Chapter 4 reviews the basic and traditional technologies of biodiesel production, and then introduces the new notable technologies developed in the past decade. Due to the major changes in biodiesel policies as stated above, various types of feedstock were newly explored and developed, and new conversion technologies were developed for more effective production of biodiesel and for converting the new feedstocks.

Chapter 5 explains the biodiesel production from green algae. Green algae has drawn much attention last decade due to its superior sustainability and production efficiency. The chapter reviews the history of algae biodiesel first, and then explains the problems of algae biodiesel and current technology development.

Chapter 6 presents the properties of biodiesel from various types of feedstock. As stated above, the biodiesel feedstock has experienced drastic change over the last decade coincided with the redirection of policies. The basic properties of biodiesel vary depending on the type of feedstock and when blended with petroleum diesel, the properties of biodiesel blends differ from the pure biodiesel depending on the blending ratio. The relationships between the properties of pure biodiesel from various feedstocks and the blends with different ratios are presented.

Chapter 7 discusses the application of life cycle assessment (LCA) in research and policy design for biodiesel. The uncertainties in evaluating the climate change impacts of biodiesel derived from oil crops, waste fats, oil and grease (FOG) and algae are critically examined.

Chapter 8 presents the future direction of biodiesel based on the current policies and the status of world economy.

This book provides unique perspectives on biodiesel from different angles. It help the readers better understand the major issues and concerns regarding biodiesel industry and economy, and prepare for the future of biodiesel in the coming decade. In summary, the readers will be able to see that the biodiesel policies are significantly influenced by energy production, energy needs, availability of feedstock, environmental issues, and tax structure of a nation, which in turn are influenced by the domestic and diplomatic political situations, and economic status of the country.

About the Editors

Dr. Hyunsoo Joo is currently a chief research fellow and director of Department of Atmosphere and Climate Change Research at Korea Environment Institute (KEI) in South Korea. His research area is air quality management, waste management and environmental impact assessment. Dr. Joo received his Bachelor of Science in chemical engineering from Seoul National University, Korea (1989), his Master of Science in environmental assessment from McGill University, Canada (2007), and his PhD in chemical engineering from the Auburn University, USA (1997).

Dr. Ashok Kumar is a distinguished university professor and chair of Civil and Environmental Engineering at The University of Toledo, Toledo, Ohio. Before coming to Toledo he worked for Syncrude Canada Ltd. as an atmospheric physicist where he developed, planned, and conducted studies related to the dispersion of emissions of a tar sands plant. Dr. Kumar received his Bachelor of Science in Engineering (Honors) from Aligarh University in India (1970), his Master of Applied Science from University of Ottawa, Canada (1972), and his PhD from the University of Waterloo, Canada (1978). He is registered as a Professional Engineer in the Province of Alberta, Canada and is a Diplomat of the American Academy of Environmental Engineers. He is a fellow member of Air and Waste Management Association (A&WMA). Air and Waste Management Association presented him with the L. A. Ripperton Award for distinguished achievement as an educator in the field of air pollution control in June 2003. He was conferred honorary membership in A&WMA in 2014. His research work has focused on finding innovative solutions to fundamental and applied problems in air quality, risk analysis, and environmental data analysis. He worked on biodiesel projects from 2004 to 2016 funded by the Department of Transportation through three different DOT university centers.

Contributors

Aashti Hamid is currently working as a consultant and is actively involved in the research areas of Green Energy, Waste Water Management and Artificial Intelligence. She is a gold medalist in the Bachelor of Technology program (Petrochemical Engineering) from AMU Aligarh, India and has worked as a Scientist Trainee while pursuing her Masters (Chemical Engineering) from NCL-CSIR Pune, India, in the field of Advance Modeling and Simulation. She has also taught various chemical engineering courses to undergraduate and graduate students at the M. S. University of Baroda, Vadodara, India.

Mohammad-Matin Hanifzadeh is a postdoctoral research fellow at the University of Houston. His research interests are in the areas of biofuel production and waste management, such as developing the methods to improve environmental/economic sustainability of the processes by manipulating the operation parameters. His degrees are a BSc (Sharif University of Technology), an MSc (University of Tehran) and a PhD (The University of Toledo) in Chemical Engineering.

Dr. Dong-Shik Kim is professor of Chemical Engineering at The University of Toledo. Since joined the faculty, Dr. Kim has worked on several research projects supported by Federal, State, and international funding agencies including NSF, NASA, USDA, Air Force, Ohio Board of Regents, and KIST (Korea Institute of Science and Technology). His research has focused on biomaterials development and biomass energy. Dr. Kim earned his Bachelor of Science and Master of Science degrees in chemical engineering at Seoul National University, Korea, and his doctorate in chemical engineering at the University of Michigan, Ann Arbor, USA. Dr. Kim is an active member of American Institute of Chemical Engineers. Recognized for his achievements in research and education, he received the deArce Memorial Endowment Award, Kohler Junior Faculty Award, and Outstanding Undergraduate Research Mentor Award. He is a recipient of Air Force Summer Faculty Research Fellowships. He has published more than 50 articles in peer-reviewed journals and presented his work in domestic and international conferences. He received Fulbright Distinguished Chair Award in 2018. He is a registered professional engineer in Michigan.

Sudheer Kumar Kuppili is a PhD Student at the Indian Institute of Technology – Madras in India. His research interests are in the areas of urban air quality, biofuels, engine performance and emissions, and emission control technologies. His degrees are a BTech in Civil Engineering (Chaitanya Engineering College), and MS in Civil Engineering (The

University of Toledo). Sudheer has also worked as an Environmental Engineer (Tetra Tech Inc.).

Hamid Omidvarborna obtained his BSc and MSc from Iran in Chemical Engineering and his PhD in Environmental Engineering (Air Quality) from The University of Toledo, Ohio (US). His research focuses on combustion chemistry of various biodiesel fuels in diesel engines at Toledo. He worked at the Sultan Qaboos University (Oman) as a postdoctoral student on a variety of environmental engineering problems, including the analysis and modelling of different ambient air pollutants. Currently, he is working as a research fellow at the University of Surrey (UK) on the implementation of low-cost sensors. His research interests are in the areas of air quality, air pollution control, environmental monitoring and assessment, energy and environment. He has published over 20 journal papers and book chapters.

Kaushik Shandilya is an environmental executive working in the State of West Virginia. His research interests include exploring different matrices of environmental engineering and sciences. He is presently focusing on the reclamation of natural environment using verified sediment and erosion control technology. He is a specialist in Article 3 and Article 11 of USEPA. His work on algae, energy efficiency, air and water quality, biodiesel and renewable energy are of wide application. His degrees are an MS (Devi Ahilya Vishwavidyalaya) in chemistry, an ME (Malaviya National Institute of Technology, Jaipur), and a PhD (The University of Toledo) in environmental engineering. He is author or coauthor of a number of peer-reviewed articles and book chapters related to air and water quality. Dr. Shandilya has taught graduate and undergraduate courses at The University of Toledo, Baker College, Owens Community College, and Govindram Seksaria Institute of Technology and Sciences.

Qingshi Tu is a postdoctoral associate in the Department of Chemical and Environmental Engineering at Yale University, where his research is to promote the development of sustainable technologies for algae biorefineries via life cycle assessment (LCA), techno-economic analysis (TEA), and other modeling approaches. Qingshi is also affiliated with the Center for Industrial Ecology at Yale University, where he is responsible for the evaluation of climate mitigation potential of resource efficiency strategies in transportation sector.

1

Introduction

Hyunsoo Joo, Ashok Kumar, and Hamid Omidvarborna

1.1 Background and Rationale

The demand for energy depends on its consumption and production of energy for any nation. There is no energy problem if the production of energy is more than the consumption in a country. However, if the reverse is true because of political or other reasons, then energy shortages would develop over time as the consumption increases and the energy supplies are tighter. The strategy for national energy security is evaluated based on the energy balance between the supply and demand. Energy gap may be minimized in several different ways in a nation. Many countries are investing heavily in research and development (R&D) and revising energy policies in order to reduce the energy consumption and to secure more energy resources. Reduction of energy consumption by using more energy efficient electrical equipment and motor vehicles is one of the ways to decrease the net energy demand. Development of renewable energy sources is another way to secure energy supplies. Policies and laws that encourage manufacturing and use of energy-efficient devices and renewable energy are also important factors that make impacts on the energy status. Intensive and interdisciplinary research is needed to achieve a positive balance both on the supply and demand sides. This report focuses on the supply side.

Many nations are concerned about greenhouse gas (GHG) emissions from transportation systems and are looking for alternatives to assure energy security of their nation. The use of natural gas, propane, alcohol, and electricity is being considered to reduce the dependence on gasoline and diesel. The terms "alternative fuels" and "renewable fuels" are used interchangeably in the literature. However, they have specific meanings in statute and regulation. The definitions even differ from law to law. The United States Environmental Protection Agency (USEPA) website points out that alternative fuels are understood to mean alternatives to traditional gasoline and diesel fuels and the renewable fuels are those derived from renewable, nonpetroleum sources such as crops, animal waste, or municipal solid waste. In this report, biodiesel is defined as a renewable energy form that is an alternative to traditional fossil fuels.

The US energy supply since 1850 is given in Figure 1.1. This figure also provides modeled projections of domestic energy markets through 2030, and it includes cases with different assumptions and methodologies regarding macroeconomic growth, world oil prices, technological progress, and energy policies. Energy market projections are subject to much uncertainty, as many of the events that shape energy markets and future developments in technologies, demographics, and resources cannot be foreseen with certainty. From the figure, one will notice that the renewable component of energy supply in the US was very small up to 2000, whereas current projection shows that its contribution will be increased to 5% in 2030. Figures 1.2 and 1.3. give a summary of energy consumption for the US and the rest of the world. Only a small portion of energy is being produced using biofuels (2.2%) in the US; while, the amount of modern renewables in global primary energy consumption is around 3%. Figures 1.1 to 1.3 show that the energy crisis during the last decades on a global scale has forced many countries to plan on diversifying the sources of energy. Instability in international relations, depletion of fossil fuel and environmental concerns were the driving forces for securing diverse energy sources.

Note that more than 80% of the world energy demand is still projected to be met by fossil fuels in 2030. To achieve energy diversification, persistent research effort and policy development for renewable energy such as biodiesel (BD, in short) are necessary. Therefore, this book focuses on in-depth analyses on the recent progress in biodiesel production technologies and trends in government policies of various countries to help policymakers

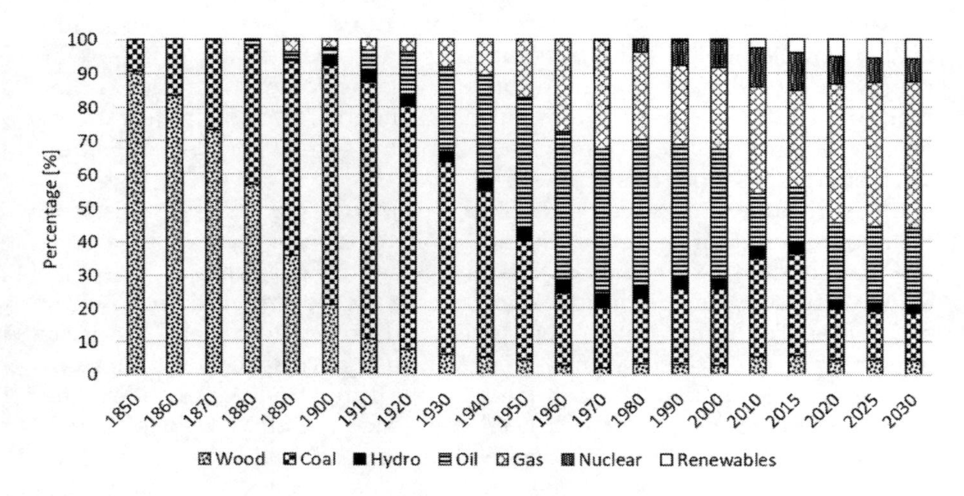

FIGURE 1.1
US energy supply since 1850. Gas includes natural gas plant liquid plus dry natural gas; biomass includes energy from wood, wood waste, and corn (Annual Energy Outlook, 2018) [1].

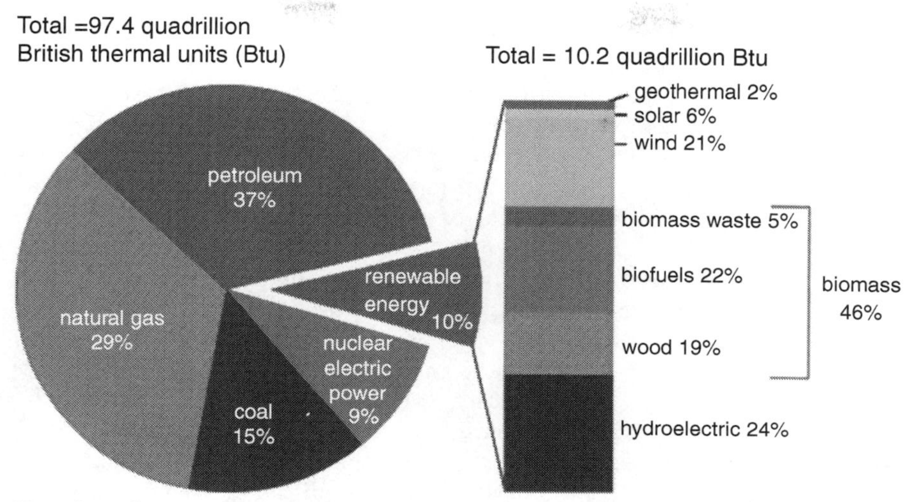

FIGURE 1.2
The US relies on fossil fuels (more than 80% of its total energy consumption). (Source: US EIA, Monthly Energy Reviews, Table 1.3 and 10.1, April 2017) [2].

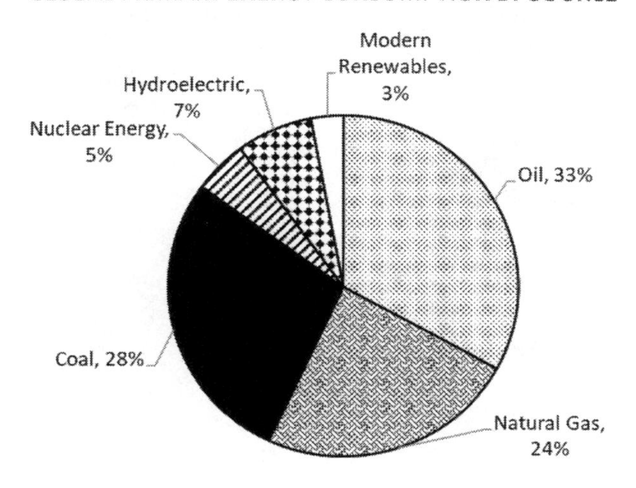

FIGURE 1.3
World energy consumption relies on oil, coal, and natural gas totally 85%. (Data Source: BP Statistical Review of World Energy, 2017) [3].

and industries make more efficient policies and plans, and to provide future direction in energy area regarding biodiesel.

Bioethanol and biodiesel made from plants instead of petroleum can be blended with or directly substituted for gasoline and diesel, respectively. Biodiesel is produced from non-petroleum-based renewable resources in many countries. Biodiesel produces less air pollution and greenhouse gases mainly due to its oxygen content. It is biodegradable, nontoxic, and safer to handle. The benefits of biodiesel on GHG emissions are significant, because CO_2, as the main compound in GHG emissions, released from biodiesel combustion is offset by the CO_2 captured by the plants from which biodiesel is produced. The use of soybean oil for producing biodiesel can offset 96% equivalent carbon dioxide (CO_2) emissions. The use of corn oil for making biodiesel can offset 39.1% equivalent CO_2 emissions. Generally, the use of biodiesel can lower the greenhouse gases by 50% compared to regular diesel. However, it is reported that biodiesel produces more nitrogen oxide (NO) and the use of blends above B80, which consists of 80% biodiesel and 20% of regular diesel is not yet warranted by automakers.

Biodiesel blends can be readily used in most diesel engines, especially in more recently produced engine models. Another concern is that biodiesel has lower fuel economy and power than that of regular diesel. Its power is 10% lower for B100 and 2% lower for B20 than regular diesel according to different studies. Biodiesel is currently more expensive than regular gasoline. B100 is generally not suitable for use in low temperatures and it drastically affects the engine durability.

Like bioethanol, biodiesel production from plant oil such as soybean oil is not considered economically feasible because it may negatively affect the food market by raising food prices and eventually biodiesel price also. Net energy expense for biodiesel production from soybean has been highly suspected by many researchers for the requirement of high energy input for production and low economic gain from the product. Biomass and waste oils are recommended for alternative feedstock for biodiesel production. Some of the most common (and/or most promising) biomass feedstocks are:

- Grains and starch crops – sugar cane, corn, wheat, sugar beets, industrial sweet potatoes, and so on
- Agricultural residues – corn stover, wheat straw, rice straw, orchard prunings, and so on
- Food waste – waste products, food processing waste, and so on
- Forestry materials – logging residues, forest thinnings, and so on
- Animal byproducts – tallow, fish oil, manure, and so on
- Energy crops – switchgrass, miscanthus, hybrid poplar, willow, algae, and so on

- Urban and suburban wastes – municipal solid wastes (MSW), lawn wastes, wastewater treatment sludge, urban wood wastes, disaster debris, trap grease, yellow grease, waste cooking oil, and so on

The world biofuel production trends from 2006 to 2016 for different continents and selected countries are shown in Figures 1.4 and 1.5 and Table 1.1. Global biofuels production rose by 2.6% in 2016, well below the 10-year average of 14.1%, but faster than in 2015 (0.4%). The US provided the largest increment (1930 thousand tonnes of oil equivalent, or ktoe). Global ethanol production increased by only 0.7%, partly due to falling production in Brazil.

The figures also show that the overall biodiesel production has been constantly increased on every continent of the globe. Biodiesel production rose by 6.5% with Indonesia providing more than half of the increment (1149 ktoe). It is obvious that the percentage of biodiesel production in Europe has been much higher than that of all the rest of the world combined as in Figure 1.4. Despite the recent decline of biodiesel production in one of the world's largest biodiesel producing countries, Germany, Europe is still the largest biodiesel producing region, and its production is still growing. It is worth to mention that the production in different countries depends on economics, tax incentives, and government policies.

The biodiesel production decreased in Germany, Italy, and the US in 2008 (Germany and Italy) and 2009 (US), which had coincided with the time when governments announced the abolition of government subsidies and tax breaks. It was also regarded that overall economy downturn in

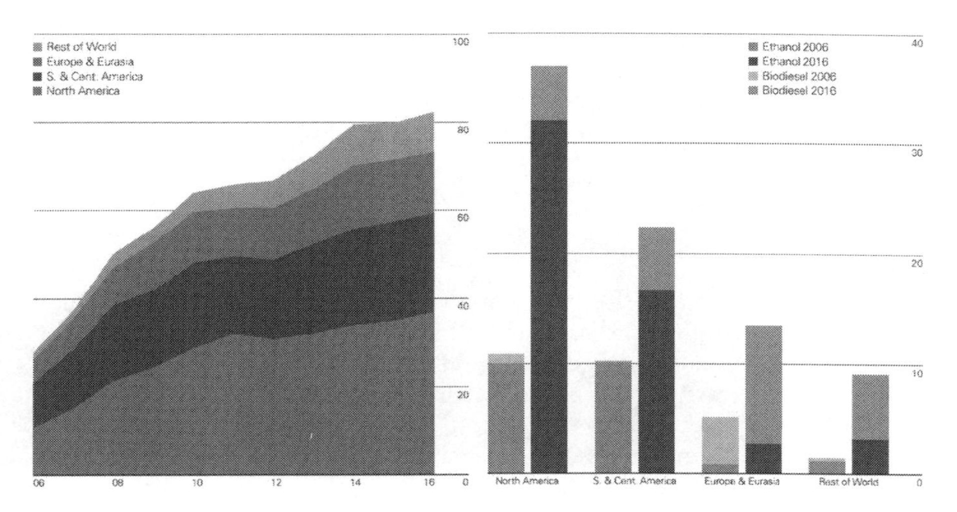

FIGURE 1.4
World biofuels production (million tonnes of oil equivalent).
(Data Source: BP Statistical Review of World Energy, 2017) [3].

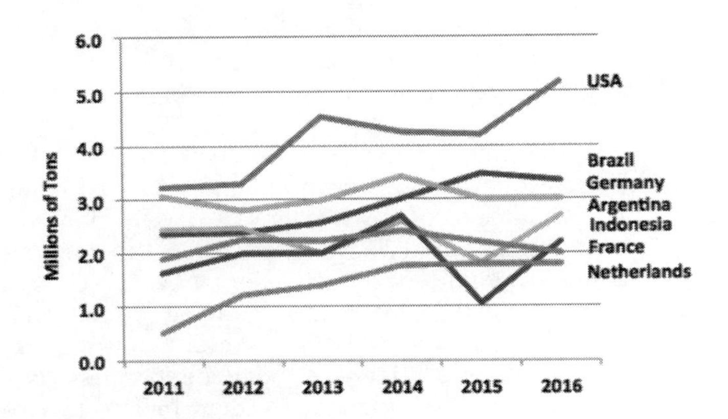

FIGURE 1.5
World biodiesel production trends from 2011–2016.
(Source: Latin American Energy Reviews, 2017) [4].

TABLE 1.1
World biodiesel production trends – country ranking – from 2011–2016.

Country Ranking in Biodiesel Production 2011–2016							2016 production
Rank	**2011**	**2012**	**2013**	**2014**	**2015**	**2016**	
1	US	US	US	US	US	US	5.2
2	Germany	Germany	Germany	Germany	Brazil	Brazil	3.3
3	Argentina	Argentina	Brazil	Brazil	Germany	Germany	3.0
4	Brazil	Brazil	France	Indonesia	France	Argentina	2.7
5	France	France	Argentina	Argentina	Argentina	Indonesia	2.2

(Source: Latin American Energy Reviews, 2017) [4].
Production in millions of tons.

the US and Europe and the stabilization of gasoline prices seemed to play a role in the reason for the decrease. However, 2016 data (Table 1.1) show that the US is by far the biggest biodiesel producer; Brazil has been second in the last two years, showing a nice rising trend, while Germany is a solid and consistent producer but whose best years may be behind them.

As shown in Figure 1.6, the US biomass energy production has steadily increased over the years. Since the early 2000s, the increasing rate has been steady and a bit higher compared to previous years. Biomass energy includes wood, waste, and biofuels, and specifically biodiesel is included in the biofuels category as well as bioethanol. In 2016, total renewable

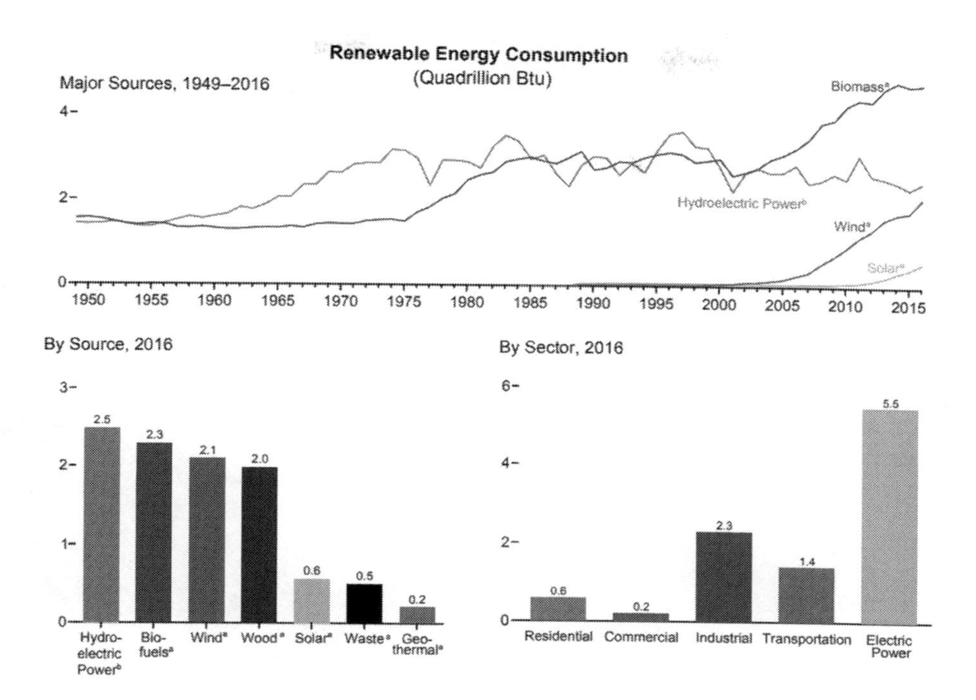

FIGURE 1.6
Renewable energy consumption in US major sources (plus sources and sectors), 1949–2016 (Monthly Energy Review, US Energy Information Administration, January 2018) [5].

energy production was 10,181 TBtu (Trillion Btu) and total renewable energy consumption was 10,112 TBtu.

Figure 1.6 also shows that among the sources, renewable energy consumption by hydroelectric power, biofuels, wind, and wood sources were much higher than other sources (solar, waste, and geothermal) in 2016. Additionally, electric power sector consumed the major portion of renewable energy (5.5 quadrillion Btu) in 2016.

The yearly trends of fuel ethanol and biodiesel production in the US are tabulated in Tables 1.2 and 1.3. It is clear that the fuel ethanol production has steadily increased since 1985 although the increasing rate slowed down a little bit in 2008.

The biodiesel production, however, decreased by about 20% from 16,145 Mbbl in 2008 to 12,281 Mbbl in 2009 and decreased about another 20% to 8,177 Mbbl in 2010. It is amazing that the biodiesel production in the US has picked up its rate in 2011, and it went back up to the amount a little higher than its highest record in 2005 despite the sluggish economic recovery. This increase is considered a very promising signal for biodiesel production in the future because it happened in the midst of tough economic recovery.

TABLE 1.2
Ethanol overview in the US.

Year	Fuel Ethanol Production Mega Barrels (Mbbl)	Fuel Ethanol Net Imports Mega Barrels (Mbbl)
1981	1,978	Not Available
1985	14,693	Not Available
1990	17,802	Not Available
1995	32,325	387
2000	38,627	116
2001	42,028	315
2002	50,956	306
2003	66,772	292
2004	81,058	3,542
2005	92,961	3,234
2006	116,294	17,408
2007	155,263	10,457
2008	221,637	12,610
2009	260,424	4,720
2010	316,617	-9,115
2011	331,646	-24,365
2012	314,714	-5,891
2013	316,493	-5,761
2014	340,781	-18,371
2015	352,553	-17,632
2016	366,981	-27,002
2017	311,982 (10-Month Total)	-24,725 (10-Month Total)

(Monthly Energy Review, US Energy Information Administration, 2017) [2].

Nevertheless, the future of biodiesel is not all that bright. According to the data in Table 1.3, biodiesel production again dropped in 2014 and 2015. Energy outlook forecast (February 2018 – as shown in Figure 1.7) show that the production of biodiesel may be reduced by 2019. Figure 1.8 also predicts the overall renewable energy production including biodiesel will be in a standstill in 2019. It is thought that overall energy production and consumption will be either staying a place or dropping due to the economic crisis starting in 2008 and in 2010 in the US and Europe, respectively. This trend last until the unemployment rate in the US and financial markets in Europe were improved. This analysis means that the increase in biodiesel production from municipal waste, used oil, and algae sources will not be enough to bump up the overall biodiesel production.

TABLE 1.3
Biodiesel overview US.

Years	Biodiesel Production Mega Barrels (Mbbl)	Biodiesel Net Imports Mega Barrels (Mbbl)
2001	204	40
2002	250	140
2003	338	-17
2004	666	-27
2005	2,162	1
2006	5,963	250
2007	11,662	-3,241
2008	16,145	-8,918
2009	12,281	-4,640
2010	8,177	-2,024
2011	23,035	-908
2012	23,588	-2,203
2013	32,368	3,477
2014	30,452	2,604
2015	30,080	6,308
2016	37,327	14,781
2017	30,875 (10-Month Total)	6,504 (10-Month Total)

(Monthly Energy Review, US Energy Information Administration, 2017) [2].

One possibility of biodiesel expansion lies in industry-driven manufacturing of biodiesel from waste oil. Along with the development of more efficient biodiesel manufacturing processes, more waste oil is expected to be collected by local business corporations and more related business activities such as sales, advertisement, and logistics, are also expected to grow. This growth is regarded as notable because it occurs by itself for economic benefits, not because of government-driven motivation and tax-benefits. The fast increase of biodiesel manufacturing from waste oil is regarded as a very strong signal for biodiesel's role in economic recovery and the much stronger growth in the near future when the economy gets better.

With this perspective, it is necessary to diversify the biodiesel feedstock. Table 1.4 shows various sources of biodiesel feedstock and their percentage of the total production of biodiesel. The items that require particular attention are highlighted. Currently, more than 69% of biodiesel is manufactured from multi-feedstock and 20% from soy. The summation of biodiesel from algae, recycled oil, used cooking oil, and waste oils, is about 1% of the total biodiesel production. Not only the overall biodiesel production is expected

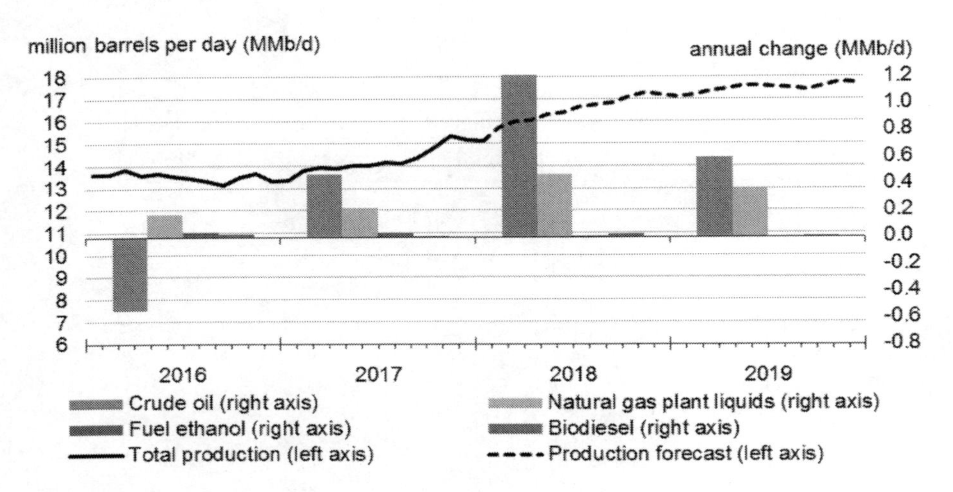

FIGURE 1.7
US crude oil and liquid fuels production
(US Energy Information Administration Forecasts, 2017) [6].

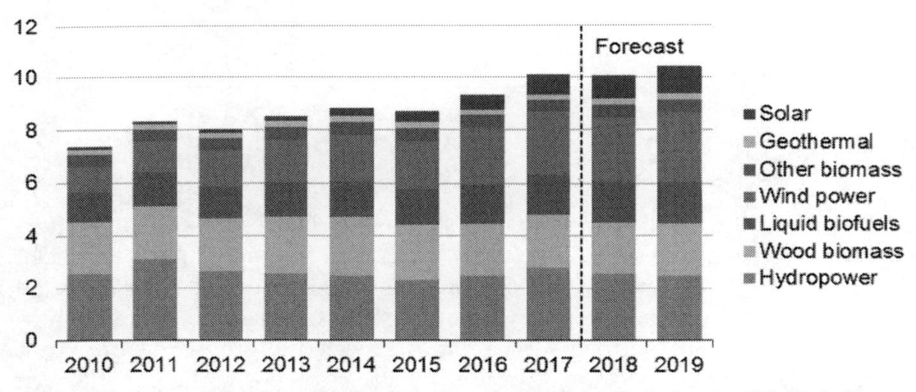

FIGURE 1.8
US renewable energy supply quadrillion British thermal units
(US Energy Information Administration Forecasts, 2017) [6].

to grow but also the portion of biodiesel from these recycled and waste oils is expected to grow. A more in-depth study is necessary to better understand the benefits and impacts of current biodiesel business depending on the feedstock.

TABLE 1.4
Biodiesel production capacity by feedstock.

Feedstock	Annual Production Capacity	% Total
Canola	93,000,000	3.87
Canola, Camelina, Safflower, Sunflower	250,000	0.01
Crude or Refined Vegetable Oils	36,000,000	1.50
Full Spectrum, including but not limited to Yellow Grease, Jatropha, Algae	3,000,000	0.13
Multi Feedstock	1,664,700,000	69.28
Palm	15,000,000	0.62
Recycled Cooking Oil	6,430,000	0.27
Recycled Cooking Oil, Tallow	1,500,000	0.06
Soy	496,900,000	20.68
Sunflower, Canola	3,000,000	0.13
Tallow	1,250,000	0.05
Used Cooking Oil	1,500,000	0.06
Waste Oil	11,030,000	0.46
Waste Vegetable Oil	2,500,000	0.10
Yellow Grease	3,000,000	0.13
Unknown	63,900,000	2.66
Total	**2,402,960,000**	100

(National Research Council, 2012) [7].

According to the literature, there is no impact of biofuel drivers on bilateral trade for the biodiesel production. Unlike many other countries, the US has less dependence on foreign energy, and because of that, research and development for renewable energy and its policies are consistent and less influenced by the fluctuation of supply and demand of energy. Therefore, this study includes R&D and policy in the US to obtain an objective assessment to current and future energy technology and policy.

1.2 Aims of the Book

In this book, strategy for energy diversification will be suggested based on the analyses of biodiesel research and development (R&D) and policy in the US and other countries. To achieve this goal, an investigation was performed in four parts:

1. Perform an assessment on government support to biodiesel R&D, production, and consumption in the US and other countries,

2. Perform a survey on the trend in biodiesel production technology and research direction in those countries,
3. Discuss the application of life cycle assessment (LCA) in research and policy design for biodiesel, and
4. Analyze the problems of biodiesel and suggest a future direction for R&D and strategy and policy for the invigorating biodiesel industry and business.

1.3 Scope and Content

The major objectives of the book are:

- To better understand the general trend, characteristics, and problems of government policies on biodiesel in the US and other countries.
- To assess the current situation of biodiesel production research and development in the US and other countries.
- Strategies for invigorating biodiesel utilization in the industry both in the US and other countries.

In order to accomplish the above objectives, the data related to biodiesel production, policy, and technology were collected for the last 10 years, 2006–2017.

Disclaimer: The views expressed in this chapter are those of the authors.

References

1. USEIA, US Energy Information Administration Annual Energy Outlook, 2018. Office of Energy Analysis, US Department of Energy, Washington, DC 20585.
2. US EIA, Monthly Energy Reviews, Table 1.3 and 10.1, April 2017. www.eia.gov/totalenergy/data/monthly/
3. BP plc, BP Statistical Review of World Energy, June 2017. www.bp.com/content/dam/bp-country/de_ch/PDF/bp-statistical-review-of-world-energy-2017-full-report.pdf
4. Latin American Energy Reviews, Intelligent debate on energy issues 2017. http://carlosstjames.com/renewable-energy/biodiesel-production-becoming-a-zero-sum-game
5. Monthly Energy Review, US Energy Information Administration, January 2018
6. US Energy Information Administration Forecasts, 2017. www.eia.gov/outlooks/steo/
7. National Research Council, 2012. Renewable fuel standard: potential economic and environmental effects of US biofuel policy. National Academy of Sciences, Washington, DC.

2

Biodiesel Production and Policy

Hamid Omidvarborna and Dong-Shik Kim

This chapter presents the characteristics of biodiesel policy in different countries, development of biodiesel policy over the past decades, analysis of general trend, challenges and issues in biodiesel policy in the United States (US) and other countries in relation to the production of biodiesel.

2.1 Current Status of Policy

2.1.1 Biofuel Policy – USA

The need of biofuels including biodiesel and their development in the US was first discussed in the Energy Policy Act of 1992 (42 USC PL 102–486). The Energy Policy Act of 2005 (42 USC PL 109–58) required 7.5 billion gallons of renewable fuel to be produced annually by 2012, and the Energy Independence and Security Act (EISA) of 2007 (42 USC PL 110–140, H.R. 6) sets targets for renewable fuels of 9 billion gallons annually for 2008, expanding to 36 billion gallons per year, or close to 1 million barrels a day (b/d), in 2015. The 2007 bill specifies that 16 billion gallons per year should come from cellulosic ethanol by 2022. The Public Law 110–343, the American Recovery and Reinvestment Act of 2009, (PL 111–5) and S.152 – Dual Fuel Vehicle Act of 2011 and S.187 – Biofuels Market Expansion Act of 2011 were the latest bills in the Senate.

President George W. Bush set out specific goals for biofuels in his State of the Union address in 2006 and 2007. He proposed a reduction of 20% in U.S. gasoline consumption in 10 years by setting a Renewable Fuel Standard (RFS) of 35 billion US gallons (130,000,000 m^3) by 2017 and by reforming and modernizing Corporate Average Fuel Economy (CAFE) standards. During his presidency, the investment in hydrogen fuels, advanced batteries, biodiesel fuels, and new fuel production methods was increased. The Obama administration also emphasized a reduction in the use of gasoline and an increase in biofuel production. There were many pending bills in the House of Representatives and the Senate including one of the most controversial regulations, US RFS in 2005 that requires refiners to blend increasing amounts of biofuels with their fuel each year.

Traditionally, biofuels policies in the US have been developed and revised mostly around corn-based ethanol due to the strong political clout of US farmers. However, the effort to reform the RFS to help oil companies reduce the cost of corn-based ethanol has acquired strength since President Trump was sworn in. The Trump administration, however, seems to be indecisive over two contradicting opinions from farmers and oil companies. It is expected that the US government under the Trump administration will try to reform the RFS law in favor of the oil companies as it promised to do so. The changes may include reducing the amount of biofuel required to be blended annually, or shifting the responsibility for blending to supply terminals. And these changes, although not so drastic, will still face severe opposition from corn-state lawmakers and ethanol producers. They will remind Mr. Trump that he also vowed to maintain the RFS during his presidential campaign in corn-belt states. Therefore, the level of reform of the RFS is expected to be limited. The aggressive reformation such as capping of the pries of blending credits, called RINs (Renewable Identification Numbers), or abolishing seasonal restrictions on high-ethanol gasoline sales, is not likely to happen due to the opposition arguing that it may undermine the economy of rural America.

For biodiesel, unlike bioethanol, there seem to be no particular changes or intention to change in US policy for a while. As stated in Chapter 1, the recent increase in biodiesel production is driven more by its economic and environmental benefits than by government mandates or subsidies. Furthermore, as the feedstock of biodiesel is shifting to nonfood sources, there is no strong political lobbying to increase or decrease biodiesel production and consumption leaving biodiesel in a more or less nonpolitical area. Biodiesel production is more influenced by the gasoline price indirectly. When the RFS is reformed, regardless of the level of reformation, the gasoline price may be lowered, and the production of biodiesel would be negatively affected.

Many states have implemented laws and programs to boost the economics of alternative fuel production and market penetration in the US. One of the notable strategies is to create demand for agriculture-based renewable energy, that is, corn-based ethanol, using state Renewable Portfolio Standards (RPS). RPS requires utilities to produce a percentage of their energy from renewable sources. Just like the RFS, RPS rules change depending on the political party in power in a state government. This partisan fluctuation in renewable energy policy is unique in the US and should be considered significant when analyzing policy making and its impact on the energy market. For example, Ohio changed its rules in 2011/12 when Republican Governor John Kasich sworn in the office replacing Democratic Governor Ted Strickland. A few examples of the change are:

1. Implementation of RPSs requires, at a minimum, that state vehicles use certain volumes or percentages of renewable fuels. Some states have taken the RPS a step further and applied it to all motor vehicles within a state.

2. Increase 12.5% Renewable Energy Resources by 2024. Under the standard, utilities must provide 25% of their retail electricity supply from alternative energy resources by 2025, with specific annual benchmarks for renewable and solar energy resources. Half of the standard can be met with "any new, retrofitted, refueled, or repowered generating facility located in Ohio," including fossil fuels, making the renewables portion of the standard 12.5% renewables by 2025.
3. The alternative compliance payment (ACP) for the renewable portion is initially set at $45/megawatt-hour (MWh) under the Strickland's legislature, but would be adjusted annually by the Public Utilities Commission of Ohio (PUCO) according to the federal Consumer Price Index, although it would never be less than $45/MWh. Likewise, the Solar Alternative Compliance Payment (SACP) was set at $450/MWh in 2009, reduced to $400/MWh in 2010 and 2011, and further reduced by $50 every two years thereafter to a minimum of $50/MWh in 2024. Compliance payments would be deposited into the Ohio Advanced Energy Fund, which provided financial support to renewable energy and energy efficiency projects within the state. Utilities may not pass along the cost of compliance payments to their customers.
4. Many rules and regulations have been created by federal agencies dealing specifically with biofuel development, creation, infrastructure, economy, efficiency, supply, and types of biofuel. The US is the world's largest energy consumer, and increasing gasoline consumption is the single most important factor behind the rising US dependence on foreign oil. To aid in reducing dependence on foreign oil, the president and US lawmakers promoted the idea that biofuels could diversify the U.S fuel supply system and reduce dependence on foreign oil. Ethanol, which is blended with gasoline, is the most widely used biofuel in the US. Government's financial support for corn-based ethanol has a long history dating back to the Energy Tax Act of 1978, which exempted fuels with at least 10% ethanol by volume from the excise tax on gasoline (US General Accounting Office, 2000) [1]. The exemption effectively subsidized ethanol by $0.40/gal. In 1980, two new options were also created, a blender's tax credit and a pure alcohol fuel credit (Solomon, Barnes, and Halvorsen, 2007) [2]. While they subsidized ethanol to the same degree, they were much less frequently used. The exemption and its equivalent subsidy stayed roughly similar in nominal terms for the next 25 years, although the benefits were allowed to go to the blends of less than 10% after the Energy Policy Act of 1992 (US General Accounting Office, 2000) [1]. The exemption and subsidies increased to $0.60/gal in the Tax Reform Act of 1984 before falling to $0.54/gal in 1990 and to $0.52/gal when they were canceled in 2004 (US General Accounting Office, 2000 [1]; Rubin, Carriquiry, and Hayes, 2008) [3].

The American Jobs Creation Act of 2004 replaced the exemption and existing credits with the Volumetric Ethanol Excise Tax Credit (VEETC) that gave the credit directly to blenders (Rubin, Carriquiry, and Hayes, 2008) [3]. VEETC replaced the previous federal ethanol excise tax incentive established by the Energy Security Act of 1979. The rate of credit was initially $0.51/gal, though it was reduced to its current level of $0.45/gal in the 2008 Farm Bill (Solomon, Barnes, and Halvorsen, 2007) [2]. VEETC expired on December 31, 2011. A cellulosic biofuel producer that is registered with the Internal Revenue Service (IRS) may be eligible for a tax incentive in the amount of up to $1.01 per gallon of cellulosic biofuel. Only qualified fuel produced in the US between January 1, 2009, and December 31, 2012, for use in the US may be eligible (Public Law 111–152, Section 1408; Public Law 110–234, Section 15321; and 26 US Code 40). As stated above, the national renewable fuels standard (RFS) program was developed to increase the volume of renewable fuel that is blended into transportation fuels, of which diesel is the most common. The EISA of 2007 set targets for renewable fuels of 36 billion gallons per year by 2022. Corn ethanol production, under the bill, is to be capped at 15 billion gallons per year. The bill specifies that 156 billion gallons per year should come from cellulosic ethanol by 2022.

Biodiesel Income Tax Credit gave an incentive of $1.00/gallon for a taxpayer who delivered unblended biodiesel, B100 (Public Law 111–312, Section 701; and 26 US Code 40A). The biodiesel tax incentive was structured in a manner that made the fuel price competitive with conventional diesel fuel in the market place. This tax credit expired on December 31, 2011. Efforts were underway in 2012 to reinstate the tax credit. The US Congressional Budget Office estimated the cost to taxpayers of reducing petroleum diesel by using biodiesel was $2.55/gallon (Gecan and Johansson, 2010) [4], which was considered a blocking block to reinstatement of the tax credit under the current economic downturn.

The Obama administration wanted 80% of electric production to come from "renewable" energy by 2035, including clean coal and efficient natural gas. Moreover, US domestic oil production was also on the rise and the expectations were that the use of Middle Eastern oil would reduce significantly by 2020. New nuclear plants, the first ones in 30 years, was approved in Georgia, and a portion of added electric production now came from wind. The 12-year plan was to reduce oil demand in the US and the 20-year plan was to increase supply alternatives. Along with it, more push for biodiesel production in the US was visible.

The Obama administration would have liked to get US autos to average 50 mpg (miles per gallon) by 2025. This goal would have quite an impact on energy demand because 70% of oil demand is from transportation. Today, the average car gets 22 mpg. The demand for oil in the US would be cut in half if the Obama administration had been successful in increasing gas consumption to 50 mpg. This tough mpg standard seems to be unattainable for a while as the Trump administration announced on April 2, 2018, that it would revise tough mpg fuel standards for cars and light trucks, saying the rules agreed to during President Obama's time in office were "not appropriate."

2.1.2 Biofuel Policy – Canada

Both the federal and provincial governments of Canada share jurisdiction over the energy policy. Thus, energy programs and incentives to support the development of biofuels can be found at both levels and often overlap in their intended purposes. The federal and provincial governments have developed different programs to assist the biofuels industry with the investment in new technology research and development (R&D), and to promote the commercialization of biofuels as a viable option for renewable energy and greenhouse gas (GHG) reduction. Policies for support from the federal and provincial governments have generally come in two forms:

1. Producer-based incentives, such as grants, subsidies, loans, tax credits, and tax exemptions
2. Consumption-based mandates, which is known as RFS for blending renewable fuels with gasoline (mainly ethanol) and diesel (main biodiesel) fuel sold in Canada.

The Canadian government launched several programs designed to promote biofuels. The *Canadian Environmental Protection Act (Bill C-33)* mandates the use of 2% biofuel in diesel fuel and heating oil by 2012 (www.ecoaction.gc.ca/newsnouvelles/20080626-eng.cfm). Canada also provides some incentives to increase the use of biodiesel in the transportation sector. The renewable fuel regulations were enacted on August 23, 2010. These regulations require 5% renewable contents (which includes biodiesel) in the gasoline pool and 2% renewable contents in the distillate pool. The *EcoEnergy for Biofuels Overview* program provides production incentive. The ethanol manufacturers are provided with incentive rates of up to CAN$0.10/L for biofuels alternatives to gasoline and CAN$0.26/L for biofuel alternatives to diesel. These incentives remained for the first three years starting from April 2008 to April 2011 of the program and then the subsidy decreased by CAN$0.04 every year until it is valued at CAN$0.06/L in 2016. In addition to this program, several other programs are in place to encourage ethanol commercial manufactures and promote biodiesel industry. These programs include ecoAGRICULTURE Biofuels Capital Initiative, Agricultural Bio-products Innovation Program (ABIP), Agri-Opportunities Program, NextGen- Biofuels Fund, Biofuels Opportunities for Producers Initiative.

2.1.3 Biofuel Policy – India

Just like the US, the policies on biofuel in India has been developed centered around bioethanol, and there aren't any policies specifically concerning biodiesel. Biodiesel production and consumption have been negligible, and there is no sizable market formed. Biodiesel will continue to grow if supported by a commercially viable strategy for building a sustainable biodiesel industry in India although the market penetration rate is expected to remain minimal for a long time. In an effort to enhance the use of biofuel, the Ministry of Petroleum and Natural Gas (MoPNG) under Government of India (GOI) through its ethanol blending program (EBP) mandated the use of 5% ethanol blend in petroleum starting in January 2003. Although much ethanol was produced from sugarcane molasses in India, this program was only partially implemented due to low sugarcane production through 2005. In 2006, a surplus sugarcane production made GOI to mandate 5% ethanol blending across the country. At that time the oil marketing companies (OMC) contracted for 1.4 billion liters of ethanol for EBP at Rs. 21.50/L ($ 0.4/L). In 2007, however, the GOI deferred its implementation due to insufficient sugarcane production. Although the National Biofuel policy approved in 2008 mandates 5% blending of petroleum with biofuel, this policy was also deferred due to the shortage of sugarcane.

The third phase of EBP envisaged the blending ratio of 10% and increasing it to 20% by 2017. As a result, India achieved its highest ever ethanol market penetration, a gasoline blend rate of 3.3% on average across the country according to the USDA GAIN report in 2016. The EBP is likely to expand, but, considering the lasting shortage of bioethanol supplies, at a slower pace as demand from industry is partly met by imports. The ethanol import grew from 500 million liters in 2017 to 600 million liters in 2018.

Currently, the GOI does not allow the use of imported ethanol for the EBP program. The GOI offers subsidized loans to sugar cane mills that cover a maximum of 40% for the creation of an ethanol production plant. Fiscal incentives for biodiesel production include the exemption from central excise tax (4%), though most state administrators have maintained the state excise duty. On the other hand, there are no direct financial tax incentives for ethanol production.

Regarding biodiesel, the GOI launched National biodiesel mission (NBM) in 2003 that identified *jatropha curcas* as the most suitable oilseed for biodiesel production. The government had initially set a target of covering 11.2 to 13.4 million hectares of land under jatropha cultivation by the end of 2012 and achieving a 10% blending target. However, at the end of 2010, only estimated 0.5 million hectares of land was under jatropha cultivation. The slow pace of jatropha planting, and the high production costs surpassing its purchasing price are effectively hindering the ambitious targets proposed by the government. The GOI also planned to fix a higher price of Rs. 34/L ($0.63/L) (compared to Rs. 26.5/L) for the purchase of biodiesel through OMC.

Currently, India has set out plans to boost its biofuels market over the next few years in order to beef up its energy security. Blending 5% of biodiesel with regular diesel and 10% ethanol with gasoline could boost the market to

$8 billion by 2022. India would require 6.75 billion liters of biodiesel and 4.5 billion liters of ethanol for blending over the six years as of 2016. More discussions are given in Chapter 3.

2.1.4 Biofuel Policy – Argentina

Argentina enacted a law (*Argentine Biofuel Law 26.093*), which requires 5% biofuel share in gasoline and diesel starting in January 2010. The manufactures can opt for reimbursement of value-added tax (VAT that isa tax charged on the sale of goods or services and is included in the price of most products and services that we use every day) or accelerated depreciation on capital investments. The manufactures can also enjoy financial support for the ethanol sold within the country and do not receive any tax incentives for exported ethanol products. However, biodiesel receives favorable export tariffs of 17.5%. In Argentina, biodiesel is produced from soybean and ethanol production is substantially less developed compared to biodiesel. The financial incentives given by the government are reviewed annually.

Argentina bioethanol production in 2018 is forecast at 1.12 billion liters. This record is the result of expected higher demand for gasoline and a real blend rate almost reaching the official cap of 12%. Biodiesel production is projected at 3.05 billion liters in 2018, the highest ever. This production forecast takes into account a modest growth in the local mandate and exports at 1.7 billion liters, unchanged from 2017. Post's forecast is based on policies in place in Argentina's main biodiesel export market through mid-July 2017. There is much speculation over the outcomes of commercial cases in the US and the European Union (EU) that could impact Argentina biodiesel exports (Joseph, 2017) [5].

The most current Policy/Regulatory Developments in Argentina is as follows: In June and July 2017 the Government of Argentina set the export tax on biodiesel at 0% (from 0.13% in May). A factor which contributed to the expansion of the local biodiesel industry since its beginnings was the differential export tax on biodiesel soybean oil. In July 2017 soybean oil exports were taxed 27% and biodiesel exports 0%. Since 2012, the Government of Argentina has in place a "flexible export tax system" for biodiesel which is modified on a monthly basis. Table 2.1 shows biodiesel, soybean oil and soybean export taxes in the past 12 months in Argentina.

In January 2017, under Decree 1343/17, the government established that the export tax on soybean and its by-products would drop by 0.5% per month beginning in January 2018 through December 2019. Therefore, the export tax on soybean oil which is currently 27% will drop to 21% in December 2018 and to 15% in December 2019.

2.1.5 Biofuel Policy – Brazil

More than any other country, Brazil has a long tradition of biofuel policy support programs, whereas the first government programs intended to

TABLE 2.1
Biodiesel, soy oil, soybean export tax of Argentina in percent.

Month	Biodiesel Export Tax in %	Soy Oil Export Tax in %	Soybean Export Tax in %
July 2017	0.00	27.0	30.0
June	0.00	27.0	30.0
May	0.13	27.0	30.0
April	6.58	27.0	30.0
March	6.02	27.0	30.0
February	5.05	27.0	30.0
January 2017	3.31	27.0	30.0
December 2016	4.98	27.0	30.0
November	5.51	27.0	30.0
October	5.27	27.0	30.0
September	5.55	27.0	30.0
August	4.99	27.0	30.0
July 2016	7.15	27.0	30.0

stimulate growth in the sugarcane ethanol industry dating back to the 1930s. From the 1930s to the mid-1970s, Brazil's Sugar and Ethanol Institute regulated different policies, for example, production quotas, price controls, and a blend mandate, to support the agricultural industry. The oil crisis in the 1970s gave the impetus to further increase and resulted in new policies on biofuels in the 1970's. Brazil introduced the *National Alcohol Program*, named *PROALCOOL*, focused on the production of ethanol from sugarcane. The objectives of this program were to limit energy supply constraints and to provide a stable demand for the excess production of sugarcane. The next decade saw major commercialization of biofuels and 96% of automobiles sold in Brazil in 1985 were ethanol powered (Colares, 2008) [6]. The legislation made in 1993 required 22% of ethanol content in the gasoline and it was increased to 25% in 2003. The invention of flex-fuel technology (FFT) made ethanol a more sought an alternative to gasoline. It was estimated that the Brazilian car fleet with FFT reaches 43% in 2015 (de Almeida, et al., 2008) [7].

There are no direct subsidies for ethanol production in Brazil. However, federal duties are much lower at US$0.01 per liter compared to US$0.26 per liter for gasoline, which give much more favor to ethanol than to gasoline. The ethanol also enjoys less fuel VAT when compared to gasoline in most of the states in Brazil. For example, in the Sao Paulo state, where most of the ethanol production is located, the VAT on ethanol is 22% while it is 47% on gasoline. It is noted, however, with low petroleum prices in recent years, ethanol's market share for flex-fuel vehicles reduced from 55% in 2008 to 30% in 2015 (Bognar et al. 2017) [8].

Based on the success of the ethanol program, Brazil initiated a program aimed at investing and developing biodiesel production. The *National Program on Biodiesel Production and Usage* (PNPB) introduced in 2005 requires replacing petroleum-based diesel with 2% biodiesel from 2008 to 2012 and increasing it to 5% from 2013 (Colares, 2008) [6]. The government mandated a blending share of 4% in July 2009 (Barros, 2009) [9]. Most of the biodiesel production is from soybean and other oil plants. The biodiesel in Brazil is subsidized through two schemes: (a) In the first scheme, the National Petroleum Agency (NAP) buys quantities of biodiesel through auctions organized by the government to ensure supply targets. The prices paid in auctions are higher than the production costs. (b) *Social Fuel Scheme* which requires a minimum percentage of raw materials to be bought from family farmers to qualify for fiscal benefits (Barros, 2009) [9].

As part of Brazilian Government's commitment to the 21st Conference of the Parties (COP21), the Ministry of Mines and Energy of Brazil proposed a regulatory framework (RenovaBio program) to revitalize the biofuels sector, encouraging energy efficiency gains in biofuels production and use, and to recognize that different biofuels have different capacities to contribute to the goals set at COP21. In August 2017, the Brazilian Government put a tariff rate quota in place for ethanol imports, allowing 600 million liters to enter duty-free, with any volume above being subject to a 20% tariff. This measure was followed by the March 2017 request by Brazilian ethanol producers to place a tariff on ethanol imports. Producers claimed the pace of imports endangered domestic ethanol production; especially in Northeastern Brazil, where import volumes rose significantly due to the competitive prices from imported corn ethanol from the US. The Brazilian ethanol-use mandate remains unchanged at 27% (E27), whereas the biodiesel mandate increased to 8% (B8) in March 2017 (Barros, 2017) [10].

2.1.6 Biodiesel Policy – Europe

- The European Union

The EU is the world's largest biodiesel producer. Biodiesel is also the most important biofuel in the EU and represents about eighty percent on an energy basis of the total transport biofuels market. Biodiesel was the first biofuel developed and used in the EU in the transportation sector in the 1990s. A rapid expansion was driven by increasing crude oil prices and the Blair House Agreement, which resulted in provisions on the production of oilseeds under Common Agricultural Policy set-aside programs, and generous tax incentives, mainly in Germany and France. EU biofuels goals set out in Directive 2003/30/EC (indicative goals) and in the RED 2009/28/EC (mandatory goals) further pushed the use of biodiesel.

In 2009 the EU approved *Directive 2009/28/EC* on renewable energy. This directive sets a minimum binding target of 10% for biofuels in transport by 2020. It also requires 35% reduction in GHG emissions for biofuels during their life cycle and this percentage increased to 50% starting from 2017. Germany and France are the two leading producers of ethanol and biodiesel in the EU. The following sections briefly describe the policies of some of the countries in the EU leading in the biofuel production.

2.1.6.1 Germany

Germany began using biodiesel in 1991 which was used in its pure form (B100). The government started encouraging the use of biodiesel in 1999 by imposing specific taxes on fossil fuel. This led to a large scale consumption of biodiesel by the end of 2007 which was equivalent to 10% of diesel consumption in the country. At the same time, the fiscal incentives to biodiesel started weighing on the country budget which reached US$3 billion in 2006. To solve these budget problems, two new legislations – Energy Taxation Law and Biofuels Quota Law – were introduced in 2006 and 2007, respectively. These legislations imposed tax on biofuels the same as fossil fuels and the discounts for B100 were progressively reduced. The discounts given to biofuels used in blends were finally abolished in 2007.

Union zur Förderung von Oel-und Proteinpflanzen (UFOP) in Germany published a report, "Biodiesel 2016/2017 – Report on Progress and Future Prospects," which provides information on the most important aspects of the national and European biofuel policy and the biodiesel markets including an overview about the legislation in the EU and Germany. Expected measures for reforming the Renewable Energy Directive (RED II) in the transport sector within the context of the EU Winter Package are presented and the national prospects for biofuels and rapeseed cultivation deriving from this package are discussed in the report. In addition, proposals for revising the RED II as well as a national policy are submitted taking into account fuels made from cultivated biomass (UFOP, 2016/2017) [11].

2.1.6.2 France

France is the leading producer of ethanol and the second largest biodiesel producer, after Germany, in the EU. France's biofuel policies include fiscal incentives and mandatory blending of bioethanol and biodiesel. The manufacturers are allocated production quotas for a period of six years through public tenders. The tax rebates are reviewed annually and have been reduced over the years. In 2005 biodiesel and bioethanol enjoyed 0.33 and 0.38 €/L ($0.5/L) reductions, respectively. The biodiesel and ethanol producers were given 0.15 €/L and 0.21 €/L in 2009. However, the fiscal rebates were phased out at the end of 2010. Table 2.2 gives a summary of incentives on biofuels in France over the years.

TABLE 2.2

Discounts* of specific taxes for biofuels in France.

	Ethanol		Biodiesel		Vegetal oil		Biofuels (2nd Generation)	
Year	**E85**	**Blend**	**B100**	**Blend**	**Puro**	**Blend**	**BTL[1]**	**Ethanol cellulose**
2004	65.05	65.05	47.04	47.04	47.04	47.04	-	-
2005	65.05	65.05	47.04	47.04	47.04	47.04	-	-
2006	65.05	0.00	38.04	32.04	47.04	47.04	-	-
2007	65.05	0.00	38.04	0.00	47.04	47.04	-	-
2008	65.05	0.00	33.64	0.00	38.89	0.00	47.04	65.05
2009	65.05	0.00	30.34	0.00	30.49	0.00	47.04	65.05
2010	65.05	0.00	24.04	0.00	22.09	0.00	47.04	65.05
2011	65.05	0.00	17.74	0.00	14.74	0.00	47.04	65.05
2012	65.05	0.00	5.14	0.00	2.14	0.00	47.04	65.05
2013	65.05	0.00	2.14	0.00	2.14	0.00	47.04	65.05
2014	65.05	0.00	2.14	0.00	2.14	0.00	47.04	65.05
2015	65.05	0.00	2.14	0.00	2.00	0.00	47.04	65.05

* All values in Euro cents per liter

[1] Biomass to liquid

(Source: Pires and Schechtman, 2010) [12]

TABLE 2.3

Energy percentage of biofuel blend quotas in France.

Year	**Gasoline+Diesel**	**Gasoline**	**Diesel**
2007	NA	1.2%	4.4%
2008	NA	2.0%	4.4%
2009	5.25	2.8%	4.4%
2010	6.25	3.6%	4.4%
2011	6.25	3.6%	4.4%
2012	6.25	3.6%	4.4%
2013	6.25	3.6%	4.4%
2014	6.25	3.6%	4.4%
	Quotas required for climate protection		
2015/2016	3% reduction in GHG emissions using biofuels		
2017/2019	4.5% reduction in GHG emissions using biofuels		
2020	7% reduction in GHG emissions using biofuels		

(Source: Pires and Schechtman, 2010) [12]

The biofuels quota law imposed mandatory biofuels sales quotas on fuel suppliers. These quotas are based on the energy content of the fuels (Table 2.3).

From 2015 onward, the biofuel quotas have been determined on the basis of GHG emission reductions. Followings are the percentage of reduction in GHG emissions using biofuels: (a) 2015–2016, 3% reduction (b) 2017–2019, 4.5% reduction (c) 2020, 7% reduction.

2.1.7 United Kingdom (UK)

The UK started promoting biofuel in 2002. Biofuel was granted a tax reduction of £0.20 per liter ($0.3/L). In 2005, this reduction was extended to E85 or blended with gasoline in any proportion. The government announced in 2008 that the reductions would end in 2010. This tax reduction policy was replaced by the Renewable Transport Fuel Obligation (RTFO) order which came into effect on April 15, 2008. The RTFO requires larger agents (those supplying more than 450,000 liters of fossil fuel per year) to sell a minimum quota of ethanol and biodiesel. Currently, the quota is 4.5% for 2012–2013 and 5% for 2013–2014 (www.dft.gov.uk/topics/sustainable/biofuels).

Following years of policy uncertainty and market stagnation, the UK Department for Transport announced its intention to raise the target under the RTFO, paving the way for the future growth of the UK biofuels industry. The government response to its consultation on the proposed amendments to the RTFO order indicated that the target level under the obligation would rise to 7.25% from April 2018, further increasing to 9.75% in 2020 and to 12.4% by 2032. By providing a long-term certainty, the department hopes to encourage future investment in a sector that investors were steadily losing faith in. The full government response on the proposed amendments to the RTFO order can be found elsewhere.

2.1.8 Biodiesel Policy – Asia

2.1.8.1 China

China's biofuel policies have focused on ethanol (grain based) production as compared to the US's biofuel policy centered on corn-based ethanol. The *Ethanol Promotion Program* was launched in 2002 to make use of excessive maize stockpiles. It was followed by the *State Scheme of Pilot Projects on Bioethanol Gasoline for Automobiles*. In 2004, the scope of the project was extended to the national level and the *State Scheme of Excessive Pilot Projects on Bioethanol Gasoline for Automobiles* (SSEP). These excessive pilot projects achieved a blending target of 10% in the selected provinces in 2006 (Dong, 2007) [13]. From 2007, the government started giving subsidies to ethanol producers in the designated provinces. A subsidy of $0.158/L was granted in 2007. From 2008 the fixed subsidy has been replaced by payments based on a yearly evaluation of individual plants' performance. The average subsidy was $0.20/L ($253/MT) in 2008, $0.19/L ($241/MT) in 2009, $0.17/L ($215/MT) in 2010, $0.16/L ($203/MT) in 2011 and $0.06/L ($79.4/MT) in 2012. The licensed ethanol producers are exempted from the 5% consumption tax and the 17% VAT. In 2009, the Chinese government announced that it would remove VAT rebate and impose 5% consumption tax on ethanol production by 2015. In 2014, China started imposing 5% consumption tax on ethanol, and in 2015, VAT exemptions for grain-based fuel ethanol producers expired. VAT exemptions for ethanol producers using "nongrain" (Generation 1.5) and cellulosic feedstocks (Generation 2.0) are still continuing as of 2017. On September 13, 2017, the Chinese state media reported on the implementation plan concerning the Expansion of Ethanol Production and Promotion for transportation fuel, jointly announced by the National Development and Reform Commission (NDRC), the National Energy Administration (NEA), the Ministry of Finance, and 12 other Ministries. The plan calls for China to achieve nationwide use of 10% ethanol (E10) by 2020. By 2025, China will shift renewable fuel production to commercial-scale cellulosic ethanol. If realized, this plan, taken with China's ongoing corn sector reform, will fundamentally transform the coarse grains, distillers' dried grains and ethanol markets in China.

For biodiesel, China has exempted the 5% consumption tax on biodiesel production using used/waste cooking oil (UCO/WCO) (main feedstock of the biodiesel production) and VAT of 17% (in 2011) over the past few years. However, consumption tax exemptions were removed in 2011. Supply and use for biodiesel remain small. There is no blending mandate nationwide, only at the provincial level at varying degrees. Along with the low support from the Chinese government, China's biodiesel industry remains in its early stages of development and faces challenges throughout the value chain. In 2015 and 2016, the market competitiveness for China's biodiesel production diminished as international crude prices plummeted, overall domestic demand for diesel fell, and quality issues plagued the domestic industry. A lack of handling standards for biodiesel feedstock and production processes led to a market decline to nearly one-quarter of its former size.

The 13th Five Year Plan includes an ambitious goal to produce 2,272 million liters of biodiesel by 2020. Previous support policies for the biodiesel industry in China energy market have been proven ineffective. 2017 biodiesel production is forecast at 500 million liters, unchanged from 2016 due to the lower overall diesel demand, limited government support, and stagnant capital investment. The decline of biodiesel production has become a visible trend as 2016 biodiesel production fell to 500 million liters, down 45 million liters from 2015, on lower overall diesel demand. Together with the domestic production decline, in 2017, biodiesel imports are forecast to slow to a trickle, unchanged from 8 million liters on weak domestic demand.

Likewise, 2017 biodiesel consumption is forecast at 488 million liters, unchanged from 2016 on continued sluggish demand. Biodiesel demand is forecast to remain flat at 521 million liters due to the slowing pace of economic expansion shrinking domestic production capacity, and declining truck kilometres-driven.

The Chinese Academy of Sciences reports that national biodiesel capacity in China is approximately 3.0 to 3.5 million tons (3,408 to 3,976 million liters). Today, China has 26 operating biodiesel plants, down five from 2015, and less than one-third of the 84 facilities operating during the industry's heyday in 2008.

One unique aspect of biodiesel in China is found in the biodiesel feedstock. UCO is the primary feedstock for biodiesel production. In 2013, researchers at Tsinghua University estimated that China is the world's leading producer of waste oil, producing 13.74 million tons of waste oil in 2010. Biodiesel production using UCO feedstock yields two primary co-products, glycerine and fatty acids, which may be further processed into industrial chemicals.

As stated earlier, there is no biodiesel blending mandate nationwide, and it is unclear if blending rates for biodiesel are enforced inconsistently, or if blending rates for biodiesel are not implemented at this time. Instead of consistent blending mandate, on December 15, 2015, China's General Administration of Customs imposed an RMB 8 per liter ($1.16) consumption tax on fuel blends with less than 30% biodiesel content. All other biodiesel blends are exempt from the measure. This consumption tax on biodiesel content on fuel is expected to boost biodiesel production and consumption to some extent. However, China's biodiesel industry is expected to remain stagnant, and to face challenges throughout the value chain.

For the other biodiesel feedstocks, there is no data to support commercially significant operations processing vegetable oil, animal fats, or advanced biodiesel feedstocks into biodiesel at this time. The share of B5 diesel (5 vol% biodiesel and 95 vol% petroleum diesel) consumption in China is difficult to measure without official diesel consumption data for B5 or other biodiesel-diesel blends.

2.1.8.2 Korea

Literature indicates that Korea is constantly adding the use of biofuels for its energy needs. The government has provided tax incentives to build biodiesel facilities. The use of biofuels in Korean energy policy is to reduce air pollution and oil dependency. This policy helps in economic growth and environmental protection. At present, Korea consumes large amounts of diesel (twice the amount of gasoline). Therefore, it is possible to add biodiesel for transportation industry by providing feedstock domestically.

Korea was producing 50 million liters of biodiesel in 2006. In 2009, it increased to 1.11 billion liters. There were 15 operating biodiesel plants with a total capacity of 625 million liters/year according to the APEC website

TABLE 2.4

Biodiesel production companies in Korea as of 2007.

Company Name	Annual Capacity (Kl)	Feedstock Type
KAYA Energy Co., Ltd.	100,000	Soybean oil and recycled cooking oil
Ecoenertech	33,000	Recycled cooking oil
Dansuk Industrial Co., Ltd.	60,000	Soybean oil
BND Energy	50,000	Soybean oil and recycled cooking oil
3M Safety	48,000	Soybean oil
BDK	33,000	Soybean oil and recycled cooking oil
Mudeung Bioenergy	6,000	Recycled cooking oil
Biodiesel Energy	9,000	Soybean oil
CNG	9,000	Recycled cooking oil
Samwoo Oil Chemical Co., Ltd.	12,000	Recycled cooking oil
Next Oil	99,000	Soybean oil
Enertech Inc.	80,000	Palm oil and Soybean oil
Bio Doil Korea	12,000	Soybean oil and recycled cooking oil
SKChemicals	34,000	Palm oil
Aekyung Petrochemical Co., Ltd.	32,000	Soybean oil
BND Kunsan	50,000	Soybean oil and recycled cooking oil
Total	667,000	

Source: The Ministry of Knowledge and Economy (MKE).

(Table 2.4), and in 2012, the number of registered companies is 23. The imported soybean oil is used to produce about 70%–80% of biodiesel and the rest is produced using UCO. Korean government provided incentives to grow rape seed in order to increase the supply of feedstock for biodiesel. B5 is available in Korea through all of its gas stations since July 2006 for passenger cars. There are about 200 stations offering B20 for fleets only. The government plans to mandate the use of B3 in Korea soon. A report by www.MarketResearch.com concluded the following:

- The South Korean biofuel consumption market had a total revenue of $593.7 million in 2011, representing a compound annual growth rate (CAGR) of 66.4% between 2007 and 2011.
- Market consumption volumes increased with a CAGR of 45% between 2007–2011, to reach a total of 2.7 million barrels in 2011.
- The performance of the market is forecast to decelerate, with an anticipated CAGR of 21.9% for the five-year period 2011–2016, which is expected to drive the market to a value of $1,597.2 million by the end of 2016.

Our analysis indicates that Korea has taken aggressive steps to introduce biodiesel to its energy mix and has been successful in producing biodiesel. More efforts could be done by legislating used oil collection and investing in technologies that encourage the use of municipal waste and farm waste. Past and current policies involving tax incentives to plant owners and farmers could be expanded to increase the supply of feedstock and to build new plants.

Table 2.5 shows the biodiesel production plan set forth in 2006 by the Ministry of Knowledge and Information (MIKE) and a list of biodiesel producers in Korea active in 2007. The number of companies has continuously increased, and as of 2012, it has increased to 23.

MKE developed a biodiesel production plan in consultation with other government agencies in 2006. The plan was focused on increasing biodiesel usage and minimizing dependence on imported feedstocks for biofuel production. As part of this plan, the current biodiesel blend ratio of 1% would be increased by 0.5% annually until it reaches 3% in 2012. Since 2006, thanks to the voluntary participation of 4 major oil companies, B20 has increased from 20 million liter in 2006 to 324 million liters in 2009.

The long-term objective is to eventually raise the blend ratio to 5%. MKE has extended industry tax breaks until at least 2010 in order to spur the needed production to meet these targets. In anticipation of increased biodiesel usage, the Ministry of Food, Agriculture, Forestry and Fisheries (MIFAFF) has made efforts to increase domestic production of biodiesel feedstocks.

In 2010, MKE held interministerial consultations to review the government's biodiesel supply plan and the blend ratio targets. The planned review looked at several key factors, including the international price of oil; the domestic and international supply and demand situation for biodiesel; and the domestic feedstock production levels.

TABLE 2.5

Korea's biodiesel mid and long term production plan.

Year	Biodiesel Blend Ratio (%)	Expected Production	
		Kiloliters	Metric Tones
2007	0.5%	90,000	79,200
2008	1.0%	180,000	158,400
2009	1.5%	270,000	237,600
2010*	2.0%	360,000	316,800
2011	2.5%	450,000	396,000
2012	3.0%	540,000	475,200
>2012	5.0%	NA	NA

* Government review of country's biodiesel plan.
Source: The Ministry of Knowledge and Economy (MKE).

In 2012, MKE launched an ambitious stipulation that is expected to boost biodiesel production and utilization:

- First, blending of biodiesel has been mandated. This legislation is set forth to compensate the effect of termination of biodiesel subsidies in 2011.
- Government will spearhead the use of animal fats and the development of feedstock farms overseas.
- Government will support the biodiesel R&D to make it a new driving force for national industrial growth.
- There will be a few years of grace period for mandating 2% biodiesel blending, and then it will increase. The level of increase will be determined by that time depending on the energy situation.
- Department of Environment will expand the collection of UCO to wider areas and more cities.
- Support for R&D has increased. A new research funding of $63.4 million has been granted for algae biodiesel research in 2012.

Starting from 2010, automotive diesel blended 2.0% biodiesel. As a result, consumption of biodiesel in South Korea reached approximately 400,000 kiloliters. The RFS program was incorporated into the act to promote the development, use and spread of biodiesel, which is called new renewable energy act by the Korean government, that was passed into law and took effect in July 2015. According to the act, B2.5 has been supplied.

Utilization of UCO was 16,000 ton in 2006 and it increased to 78,000 ton a year in 2009. This collection of used oil was mostly from commercial sectors (restaurants and food companies), and more than 60% of used oil in these sectors are currently collected and used for biodiesel production. Used oil collection from household is still an infant stage (15,000 tons), but it will be systematically increased.

Government stipulated a code for supporting overseas feedstock development business in 2012. Thanks to this support, 110,000 hectares and 30,000 hectares of jatropha and palm tree farm have been developed in foreign countries. Currently, an additional investment of $14.4 million has been made to develop 134,000 hectare farm for biodiesel feedstock in Southeast Asian country.

2.2 Biodiesel Policy Changes over Time

Changes in biodiesel policy appear to be mostly dependent on the energy prices and economic status of the country. It is also determined by the philosophy of the government on renewable energy and global climate change. The notable shift, over the last few years, in vegetable oil values to a higher plateau and the tendency, in a number of countries, to curtail direct government support for biofuel producers has posed new challenges for vegetable oil-based biodiesel producers worldwide. In the mid-2000s, during the rapid build-up in biodiesel capacity, production facilities were typically set up at low capital cost, featuring only basic refining options. Therefore, in order to remain competitive, such units are under pressure to undertake additional investments that would allow them to, inter alia, process cheaper and lower-quality feedstock, keep in storage bigger amounts of feedstock, produce higher-quality biodiesel, or recycle and sell by-products. Adjustments that allow processing of lower-quality feedstock – such as animal fats or recycled oils, which are characterized by higher levels of free fatty acids – seem to be of particular interest to the industry. The following table (Table 2.6) presents the summary of major changes that took place since 2008 in selected countries.

TABLE 2.6

Biodiesel policy change in selected countries.

Year	Notable Policy Developments
2008	**EU** – Common agricultural policy to schedule the abolishment of the energy crop premium of € 45/ha in 2010, which applies to rapeseed grown for biodiesel production. This policy brought in a negative impact on biodiesel production. **THAILAND** – While raising the reference price at which farmers sell palm fruit to crushers to prevent prices from falling and to support palm tree farmers, the use of palm as biodiesel feedstock is regarded uneconomical, which leads to an excess of supply over demand.
2009	**EU** **January** – The EU decided to retain its binding target to have 10% of its vehicle fuels sourced with renewable fuels by 2020. To encourage the use of nonfood feedstock, second-generation biofuels will count double against the target, while feedstock grown in high biodiversity and carbon stock areas will not be admitted. Specific sub-targets according to the type of feedstock used and the form of transport have not been set. By 2017, in order to be counted against the target, biofuels need to achieve at least 50% reduction in GHG emission over conventional fuel. For the time being, the carbon footprint of related ILUCs will not be taken into account. A further review of these policies is envisaged for 2014. **March** – From mid-March, imports of biodiesel from the US will face temporary antidumping and antisubsidy duties. Tariffs ranging from 26 to 41 € per 100 kg are expected to apply for an initial period of six months. Surging imports of biodiesel produced in the US – and benefiting from government support there – are considered to have severely injured the competing EU biodiesel industry. **April** – The Commission has approved the importation and processing (but not cultivation) of a GM rapeseed variety for food, feed and other uses for a period of 10 years. The new variety of T45 is characterized by tolerance to a particular herbicide and is grown primarily in Canada, whose exports toward the EU are expected to resume thanks to the approval. The impact on biodiesel production from this policy is not expected to high because, although

(Continued)

TABLE 2.6 (Cont.)

Year	Notable Policy Developments

the feedstock price for biodiesel may lower, the reduced subsidies and tax breaks on biodiesel production cannot overcome the benefits of the lowered feedstock prices.
June – An official report states that the EU is unlikely to reach the indicative 2010 target of 5.75% of transport fuel coming from renewable energy sources. A community-wide share of 4% is expected instead. The respective share in the year 2005 was 1%. Biodiesel accounts for roughly three-quarters of renewable fuel use; bioethanol contributes 15%. The EU Commission estimates that by 2020, some 20–30% of the utilization target is going to be met by second-generation biofuels.

UNITED STATES
January – Toward the end of 2008, the use of animal fat as feedstock for biodiesel production has dropped sharply in response to the introduction of stricter biodiesel standards and as prices for vegetable oils abandoned their record level.
June – An expanded RFS-2 issued in draft form by the Environmental Protection Agency (EPA) has attracted criticism from the industry. In assessing the life cycle GHG impact of biofuels, the standard considers direct and indirect emissions, including those related to ILUCs. The justification and methodology for attributing emissions from land conversion to soybean-based diesel as well as the practicability of the required feedstock certification are being questioned. Achieving the country's biomass-based diesel targets under the proposed RFS-2 would be challenging if soybean-based diesel was excluded.

GERMANY
February – In line with the overall policy to gradually scale back support provided to the biodiesel industry, from January 2009 onward, taxation of vegetable oils used as biodiesel feedstock will increase from 0.10 to 0.18 € per liter, while the tax for B100 is raised from 0.15 to 0.18 € per liter. This policy is expected to decrease the production of vegetable oil-based biodiesel and to spur the use of non-food-based biodiesel.
March – The EU Commission has requested that the sustainability criteria included in national law on the promotion of biofuels be streamlined with the relevant EU directives. As a result, importation of vegetable oils – especially soy and palm oil – for local biodiesel production is expected to resume. Furthermore, new legislation excluding biofuels that have previously received state aids from being eligible for national incentives and mandates will go ahead. Consequently, the importation of biodiesel that has benefited from state aid could become uneconomical.

FRANCE
March – Similar to Germany, where taxes on biodiesel and biodiesel feedstock have been raised in January, France has announced that all tax advantages granted to biofuel will be discontinued by 2012.

INDONESIA
February – The government is reported to consider subsidizing the sale of palm oil-based diesel when low fossil fuel prices compromise the profitability of biodiesel production.
March – The government confirmed its plan to subsidize the sale of biofuels by state-owned companies depending on how the prices for fossil fuel and biofuel feedstock (notably palm oil) develop. The subsidy would amount to Rp. 1000 per liter of biofuel distributed. Under the currently envisaged level of government funding, the sale of up to 750,000 tons of biofuel could be supported in 2009 – an amount close to the country's

(*Continued*)

TABLE 2.6 (Cont.)

Year	Notable Policy Developments
	mandatory consumption of biofuels in 2009. Palm oil-based biodiesel should account for about three-quarters of all biofuel sales.
	MALAYSIA
	March – Government sources confirmed that blending of diesel fuel with 5% palm oil biofuel will become mandatory in January 2010. In 2009, all government vehicles have started using the blend. To remain competitive, biodiesel sales shall benefit from a subsidy. Once fully operational the program is expected to absorb 500,000 tons of palm oil annually.
	April – Government efforts to prevent oversupply on the domestic market continue. Under the 200,000 ha oil palm replanting scheme, by March, 63,000 ha were approved, implying a temporary reduction in palm oil output by approx. 220,000 tons. Mandatory use of palm oil-based B5-diesel in the entire transportation and industry sector is set to start January 2010, entailing a market offtake of approximately 500,000 tons of palm oil. The increase in the country's duty-free export quota for crude palm oil to 3 million tons in 2009 is expected to stimulate exports.
	BRAZIL
	May – From January 2008, all diesel sales had to include 2% of biodiesel, which created a market for about 800,000 tons of biodiesel. Biodiesel production capacity is reported to have climbed to 2.6 million tons in 2008. In order to better use this capacity, the mandatory admixture was raised to 3% in July 2008, and currently, a further increase to 4% is under consideration. The obligatory 5% blend envisaged for 2013 has been anticipated to 2010. The main feedstock used for biodiesel production is soybean oil; its share in total feedstock use was close to 80% in 2008.
	September – Effective July 2009, the mandatory content of biodiesel in diesel is 4%; it started at 2% in Jan 2008 and became 3% in July 2008. A further increase to 5% is expected for January 2010, or three years earlier than originally planned. Annual biodiesel consumption is estimated to exceed 1.5 million tons by the end of this year.
	THE UNITED STATES and THE EUROPEAN UNION
	May – In the US, soybean-based biodiesel has been officially estimated to reduce GHG emissions by 22% over conventional diesel. This compares to a minimum 50% requirement for biodiesel to qualify toward the annual consumption targets. Other feedstock with a lower footprint, for example, animal fats or waste grease, might be required to meet the mandatory biodiesel target set for 2012. In the EU, the default values for GHG emission savings have been set at 31% for soy-diesel, 38% for rapeseed-diesel and 19% for palm oil-diesel, which compares to a minimum savings requirement of 35% (from 2013 onward). Biodiesel producers may claim greater GHG emission savings, but the onus of proof will be entirely on them.
	NEW ZEALAND
	June – To encourage investment into national biodiesel production, from July and for a period of three years, a subsidy will be granted on sales of domestically produced biodiesel that complies with sustainability standards set by the government. B5 blends are widely accepted in the country, but mandatory consumption targets have been withdrawn in December 2008. All imported material needs to meet the same standards.

(*Continued*)

TABLE 2.6 (Cont.)

Year	Notable Policy Developments
	In 2009, several countries reported about their plans regarding biodiesel utilization mandates: Coming November, in Canada, Manitoba will be the first province to introduce a mandatory blend (B2). In Uruguay, the introduction of B2 was postponed from January to October 2009 due to delays in setting up the fuel distribution system. Brazil confirmed its plan to move from B4 to B5 in July 2010, that is, three years ahead of the originally envisaged date; the higher blending level is estimated to push annual biodiesel demand to 2.2 mill tons. In Argentina, mandatory biodiesel use is confirmed to start in January 2010 at the B5 level, implying an annual biodiesel requirement of about 700 thousand tons. The government of Israel has put programs for the gradual introduction (over the 2009–2011 period) of mandatory blending (B5) on hold along with the planned tax breaks. Colombia set to start mandatory biodiesel blending of 10% from January 2010 – up from 5% in 2009. The new target is estimated to translate into an annual requirement of 500,000 tons of biodiesel – entirely based on domestic palm oil, production of which is set at 0.8–1 million tons in 2010.
2010	**EUROPEAN UNION** **January** – Norway decided to significantly reduce the tax break enjoyed by biodiesel; the measures are taking effect in January 2010. Similarly, Italy plans to drastically cut the biodiesel quota that enjoys reduced excise duty. By contrast, in Germany, the government decided to postpone a gradual rise in taxes levied on biofuels that was to be implemented from 2010 onward.
	UNITED STATES **March** – The new program (referred to as RFS-2) that came into effect in February sets the target for national consumption of biomass-based diesel in 2010 at 0.65 billion gallons, corresponding to 1.1% of total diesel sales, to which the 0.5 billion gallons that were not consumed in 2009 can be added. Reportedly, 20% of the target may be carried into 2011. The mandate will increase gradually in 2011 and 2012, while subsequent targets will be determined later. Furthermore, starting this year, measurement of GHG emissions during fuel and feedstock production, distribution and use will include emission stemming from ILUCs. Biomass-based biodiesel must lead to at least 50% reduction in lifecycle GHG emissions compared to petroleum-based diesel. According to the new RFS-2 standards, soy oil and waste oils, fats, and greases all comply with this requirement. Reportedly, the default GHG emission savings of soy oil-based biodiesel has been set at 57%, which, depending on the production method, can even reach 85%. For domestically produced feedstock, no additional certification will be required. As to competing for palm oil, it may or may not meet the requirements depending on how it is produced; compliance will be determined based on certification provided by the producer. Overall, under RFS-2, demand for soybean as biodiesel feedstock can be expected to rise in the long run, likely lifting domestic soybean prices (and hence farmer incomes as well as food costs) and reducing the country's soybean exports compared to the 2006 reference situation. **April** – The tax credit of $1 per gallon of marketed biodiesel – which expired in December 2009, negatively affecting the profitability of biodiesel production – has been retroactively renewed until the end of 2010. With the renewal, demand for soybean oil

(*Continued*)

TABLE 2.6 (Cont.)

Year	Notable Policy Developments
	and other feedstock for biodiesel is set to resume. Currently, around 10% of annual soy oil output is absorbed by the biofuel industry. **CANADA** **October** – As part of the Ecoenergy for Biofuels Programme, the government is set to provide support to a private producer of biodiesel and glycerol who uses animal fat and WCO as feedstock. The end-product shall be sold on the domestic market as well as exported to the US. The investment is expected to help reducing GHG emissions while promoting a sustainable environment. **SOUTH KOREA** **January** – Reportedly, the government has decided to raise mandatory biodiesel blending to 2% in 2010 compared to the 2009 level of 1.5%. In addition, also the tax due on diesel fuel will be waived. Korea started blending with biodiesel in 2006 as the first country in Asia. **THAILAND** **December** – Reportedly, the shift from mandatory use of B3 to B5 will be delayed from mid- to late January 2011, as flooding has adversely affected local palm oil production. **AFRICA** **July** – **BOTSWANA**, reportedly, is set to join the list of African producers of jatropha-based biodiesel. Preparations are underway to build a plant with an annual capacity of 50 million liters. Before sufficient quantities of jatropha oil become available, tallow and UCO will be the main feedstock. Under the government-backed project, initially, 170,000 ha of land will be planted with jatropha. Once developed, plantations are supposed to be leased to private companies as well as to local farmers. Biodiesel production aims at both the domestic and export market. July – **MOZAMBIQUE**, reportedly, envisages drafting its own set of sustainability criteria for biofuel production – as opposed to criteria formulated by policymakers or the industry in more developed nations. The objective is to address the country's specific needs before meeting the requirements of others. Food security considerations, as well as the reduction of expensive and volatile petroleum imports, will likely feature high on the list of criteria, compared to, for example, biodiversity preservation. According to independent experts, the main challenge will be to ensure that sustainability schemes and national policies result in the adoption of best production and management practices by the private sector and NGOs. **INDONESIA** **July** – The country's 2011 budget includes provisions for continued support to biofuel production. Sales of biodiesel and bioethanol will be subsidized at a rate of Rp. 2000–2500 per liter whenever their price exceeds the market price for mineral oil-based fuel. In the years 2009 and 2010, biofuel producers were compensated at a rate of, respectively, Rp. 1000 and Rp. 2000 per liter. Currently, marketed diesel contains 5% of palm oil-based biodiesel. **ARGENTINA** **August** – The government announced the imminent issuance of a resolution that increases the mandatory biodiesel blending rate from 5% to 7%. Furthermore, the intention to lift the rate further to 10% by the end of this year was confirmed. The country

(*Continued*)

TABLE 2.6 (Cont.)

Year	Notable Policy Developments
	is the world's leading exporter of soyoil – Argentina's primary feedstock for biodiesel. Expanding local demand for biodiesel may eventually reduce export availabilities. Reportedly, the measure is meant to provide support to domestic soyoil sales and reduce dependence on soyoil exports as well as to reduce outlays on fossil fuel imports.
2011	**CHINA** **January** – Biodiesel derived from waste animal fats or vegetable oils has been exempted from paying consumption taxes. The exemption applies retroactively from January 2009 and is estimated to amount to around 900 Yuan per ton. The measure is part of government efforts to improve competitiveness in the bioenergy sector. Reportedly, national biodiesel use is targeted at 2 million tons in 2020. **August** – Two national standards defining the quality of specific biodiesel blends have been issued in the last few years. Now a complete trade standard for biodiesel has been officially launched. It is expected to comprehensively regulate the country's biodiesel industry. Production, distribution and sales of biodiesel will be regulated in detail. The standard includes mandatory waste treatment and by-product recycling. Reportedly, a biodiesel company that already produces according to the standard said that it expects better client confidence and increased sales. Currently, the country's sales of biodiesel as a transportation fuel are reported to be minimal due to small profit margins. So far, biodiesel use (B5 blends) is mandatory only in Hainan Province.
	UNITED STATES **January** – Finally, the extension of the biodiesel tax credit ($1 per gallon) has been approved – referring to the year 2011 as well as, retroactively, to 2010. As a result, biodiesel production is expected to quickly return to previous levels. **October** – After determining in 2010, that rapeseed oil-biodiesel meets the country's requirements for biomass-based diesel, the EPA has officially approved the use of Canadian rapeseed oil in US biodiesel production and consumption. The decision opens the US market to imports of biofuel feedstock or biofuel from Canada. **December** – Reportedly, the USDA announced that it will grant US$44.6 mill to some 150 biofuel producers in the US to help support their production and product development efforts. The subsidy will be allocated based on the amount of biofuel a company produces from renewable biomass.
	CANADA **March** – Compulsory blending of conventional diesel with 2% biodiesel has come into effect across the nation in February.
	AUSTRALIA **May** – The country introduced anti-dumping duties on imports of biodiesel from the US. Triggered by complaints from local biodiesel producers, an official investigation found that shipments of subsidised biodiesel had been sold to Australia, harming the domestic industry. Australia itself currently offers a tax break worth around A$0.40 per liter to producers and importers of biodiesel. The measures are scheduled to expire in June this year. Australia's antidumping initiative is very similar to measures introduced in the EU.

(*Continued*)

TABLE 2.6 (Cont.)

Year	Notable Policy Developments
	MOZAMBIQUE **August** – Private sources reported that the government passed into law a regulation mandating nationwide use of B3 blends by the year 2012. The development of a domestic biofuel market has been promoted since 2009 and a considerable number of biofuel projects have been established in the country during the last few years.
	MALAYSIA **December** – Launched in June 2011, mandatory sales of B5 biodiesel in the three central states of Peninsular Malaysia are reported to proceed as planned. A total of 890,000 tons is expected to be required to satisfy demand (based on an estimated annual diesel consumption of 2.49 billion liters). The program relies on public subsidies to the tune of RM 106 mill per year. The subsidy is required to maintain the price of biodiesel at RM 1.80 per liter, the same as regular diesel. At present, the subsidy provided per liter amounts to RM 0.0426. Nationwide introduction of B5 is scheduled for early 2013.
2012	**EUROPEAN UNION** **January** – In Germany, biodiesel produced from UCO – even when the latter includes animal fats – now qualifies for double counting against the EU's consumption targets for renewable energy. Such regulation already applies in the Netherlands, the UK, and France, while Italy and Spain are expected to soon follow suit. The policy allows countries to use a wider range of feedstock for biodiesel production and is said to help stabilizing the price of biofuel feedstock. **March** – It is reported that less than 50% capacity utilization on average in 2011 has led to factory closures and industry consolidation across the EU. Last year, biodiesel production is reported to have fallen after a decade of rapid expansion. Main reasons behind the drop are competitively priced biodiesel imports, firm feedstock (vegetable oil) prices and weaker political support due to concerns about the environmental benefits of biofuels. Industry experts expect the challenges encountered last year to persist in 2012. **November** – The European Commission has proposed capping at 5% the contribution that first-generation biofuels (i.e., those produced from oilseeds, starch crops or sugar) are allowed to make toward the EU transport fuel consumption target for 2020. The existing EU target envisages that as much as 10% of road and rail energy come from renewable sources – a requirement that, based on new findings, could prove insufficient in the fight against global warming: recently, doubts have been expressed regarding the GHG reduction potential of several crop-based biofuels, notably when the impact of ILUCs is factored in. Reportedly, under the newly proposed target, EU production of crop-based biofuels would not be allowed to grow beyond the current level. The Commission's proposal would also require reporting of estimated emission levels for each biofuel – including ILUC factors. However, the latter requirement would remain void of legal implications, in that all biofuels would continue to be treated equal, at least until 2020. From that year onward, though, only fuels meeting strict GHG saving thresholds – including ILUC factors – will qualify, which could lead to the exclusion of biofuels produced in less sustainable ways. In this respect, depending on the criteria used, several oil crop-based fuels could be disqualified. Finally, the Commission's proposal also includes plans to remove all subsidies for growing biofuel crops. To enter into force the draft proposal will require approval by all EU member governments as well as by the European Parliament.

(*Continued*)

TABLE 2.6 (Cont.)

Year	Notable Policy Developments
	UNITED STATES
	January – it remains to be seen whether or not the respective tax credit is going to be renewed for biodiesel, while the protective import tariffs and a tax credit for bioethanol producers have been discontinued at the beginning of 2012. The fiscal incentive formally expired on December 31. Thanks to its reinstatement last year – following a one-year lapse in 2010 – the biodiesel industry has set a production record in 2011. Meanwhile, the Illinois state government is said to have extended a local tax credit (in force since 2004 and due to expire end of 2013) by 5 years. Illinois is said to have become the nation's leading biodiesel producing state. Overall, changes in federal biofuel support policies have major repercussions for the energy and food industries and markets within the US as well as internationally. As to the world market, a slow-down in global biofuel production growth has been reported for 2011, which experts attribute to poor profit margins (reflecting costlier feedstock) and lower investment in the sector.
	May – The federal government has launched two new grant schemes (together worth almost $80 million), one giving assistance to existing biofuel operations and the other one reserved for companies that conduct research on advanced biofuels, bioenergy and bio-based products.
	September – Up to 125 biofuel companies – mostly producers of biodiesel – will be awarded public funds under the USDA's "Bioenergy Program for Advanced Biofuels". The program grants payments to eligible producers based on the amount of biofuels that are produced from renewable biomass, other than corn starch. Vegetable oils and animal fats qualify as feedstock for advanced biofuel.
	November – EPA decided to raise the amount of bio-based diesel required to be included in fuel markets in 2013 to 4.28 million tons, which compares to a mandatory consumption target of 3.34 million tons in 2012. The term "bio-based diesel products" refers to biofuels derived primarily from vegetable oils, animal fats and waste oils (also considered as "advanced biofuels"). Considering the prospective slowdown in soybean oil production and supplies during the 2012/13 marketing season, market observers believe that the higher biodiesel target could raise the country's import requirements, notably of palm and rapeseed oil.
	INDONESIA
	February – Reportedly, for the fiscal year 2012, the subsidy granted to biodiesel manufacturers has been raised further to Rp. 2500–3000 per liter distributed (compared to, respectively, Rp. 1000, 2000, and 2500 in fiscal years 2009, 2010 and 2011).
	May – Starting July of this year, locally produced biodiesel must account for 2% of the total fuel consumed by the country's mining industry. The measure aims at reducing pollution and promoting domestic use of renewable fuels. Reportedly, several companies already agreed to comply with the new requirement.
	SPAIN
	May – Spain announced that petroleum companies may only use EU-produced biodiesel to comply with the country's mandatory blending requirements. Though not banning biodiesel imports from non-EU nations (notably Argentina and Indonesia), the measure is likely to make such imports unattractive while conferring an advantage to biodiesel producers in the EU, and especially in Spain. Reportedly, over the last few years, sharply rising biodiesel imports from overseas resulted in Spain's biodiesel industry working at less than 15% of installed capacity.

(*Continued*)

TABLE 2.6 (Cont.)

Year	Notable Policy Developments
	AUSTRALIA **June** – Although no comprehensive biofuel regulations, consumption targets or mandatory blending rates are in place at the federal level, biofuel companies are eligible for tax exemptions and benefited from state financial support, notably in New South Wales and Queensland. Unofficial sources report that, recently, also the government of Victoria started supporting the biofuel industry by co-financing the construction of a biodiesel blending facility.
	ARGENTINA **September** – Effective August 2012, the export tax on soybean oil-based biodiesel will amount to 32%, the same rate that applies to shipments of soybean oil. The former, lower rate of 20% (in place since 2008) stimulated the country's exportation of biodiesel, which climbed from less than 0.2 mill tons in 2007 to around 1.7 mill tons in 2011, making Argentina the world's largest biodiesel exporter. Reportedly the measure aims at making biodiesel more affordable in the domestic market, given that Argentines are paying more for the fuel than foreign buyers. The tax increase should slow down biodiesel exports (which currently absorb up to 70% of domestic soy oil production), possibly benefitting soybean oil shipments. To encourage domestic demand, along with the export tax increase, the government also lowered the official national price for biodiesel by 15% (from Ps 5196 per ton to Ps 4405). For 2012, biodiesel production has been estimated at around 2.7 mill tons, while domestic consumption should be just under 1 mill tons – slightly less than the theoretical requirement resulting from the present mandatory blending rate of 7%. To what extent local biodiesel consumption will grow, possibly compensating for lower exports, remains uncertain. Moving to a higher mandatory blending rate of 10% would lift domestic demand, but no firm date has been provided for such a step. **November** – After introducing major changes to its policy on biodiesel exports and domestic sales in August, the government decided to make further adjustments in September: initially ramped up from 20% to 35%, the country's export tax has been reduced again (first to 24.2% and then to 19.1%), whereas the government-set internal price for biodiesel has been lifted beyond the level prevailing prior to August. The adjustments were made because the combination of higher taxes and lower domestic prices, coupled with a drop in international biodiesel quotations, had led to heavy losses among domestic biodiesel producers. From now on, the tax rate and the local sales price are supposed to be reviewed every 15 days – taking into account the development of international prices for soy oil and biodiesel. The ultimate objective remains to stimulate domestic biodiesel consumption while at the same time guaranteeing adequate margins for processors.
	NEW ZEALAND **November** – A government program launched in 2009 to encourage investment into national biodiesel production – comprising a subsidy on the sale of domestically produced biodiesel – has expired without being renewed. Due to limited industry investments, take-up of the subsidy is reported to have been slow. According to market observers, investors expected more support as well as long-term assurances from the government. As to mandatory blending, the government stopped such requirement back in 2008. To date, much of the country's demand for biodiesel has been covered by imports.

(Continued)

TABLE 2.6 (Cont.)

Year	Notable Policy Developments
	THAILAND **November** – The Thai government decided to raise the country's mandatory blending rate to 5% (B5), up from 4% in effect since late 2011. Originally planned for end 2011, the government preferred to delay B5 implementation several times due to seasonal shortfalls in domestic palm oil production that led to shortages in edible oil supplies. Finally, B5 has been approved, based on improved and stable supplies of crude palm oil, which remains the principal biodiesel feedstock.
2013	**ARGENTINA/SPAIN** **January** – According to official sources from Argentina, Spain has lifted its requirement that petroleum companies may only use EU-produced biodiesel to comply with the country's mandatory blending requirements. Allegedly, though not an outright import ban, Spain's requirement severely curtailed access of foreign suppliers (notably from Argentina) to the Spanish market. Spain is said to have revoked its policy after Argentina filed a formal complaint at the World Trade Organization (WTO). **December** (Argentina) – The government decided to raise mandatory blending of diesel transportation fuel with soy oil-based biodiesel from formerly 8% to 10%. The new rate will apply from January 1, 2014, and also concerns fuel burned for power generation. The increase was introduced to compensate domestic biodiesel producers for European trade lost after the EU imposed strict anti-dumping duties on biodiesel imports from Argentina earlier this year. Once fully implemented, the higher rate is estimated to raise annual domestic biodiesel consumption by 450,000 tons.
	UNITED STATES **January** – US lawmakers decided to revive the $1.00 per gallon tax credit for biodiesel, which expired back in December 2011. The incentive, which is now scheduled to expire at the end of 2013, will be applied retroactively to all of 2012, thus granting a windfall profit to the nation's biofuel producers. Along with the revival of the biodiesel subsidy, the tax credit for biomass-based diesel, as well as a special tax break granted to small agriprocessors producing biodiesel, has been extended. Furthermore, from this year also algae-based biodiesel is eligible for support. Biodiesel producers and soy farmers welcomed the extension of the subsidy. In recent years, roughly 1/4 of total US soy oil output has been devoted to biodiesel production. Critics of public biofuel subsidies, on the other hand, pointed at the high economic and environmental costs associated with such programs, questioning the effectiveness of such measures as a means to address climate change concerns. The problem is not unique to the US. In the EU, the existing biofuel support schemes are currently being revisited due to concerns about their effectiveness and overall impact at the local and global level. **April** – The US EPA approved camelina oil as a biodiesel feedstock. According to EPA's evaluation, camelina oil-based diesel meets the 50% GHG reduction threshold required to qualify as "biomass-based diesel" or "advanced fuel" under the US bioenergy policy. Possible uses include transportation fuel, jet fuel as well as heating oil. In recent years, camelina oil has been extensively studied by the US military as a biofuel blendstock. **June** – The EPA informed that it accepted a petition from Canada by which all biofuel and biofuel feedstock, including rapeseed, approved in Canada automatically meet the requirements of the US RFS program. EPA's approval opens the doors for the exportation of Canadian rapeseed to the US for use in biodiesel production.

(*Continued*)

TABLE 2.6 (Cont.)

Year	Notable Policy Developments
	EUROPEAN UNION **January** – On January 30, the European Commission introduced EU-wide mandatory registration of biodiesel imports originating from Argentina and Indonesia. Related to the bloc's ongoing antidumping and antisubsidy investigations, the initiative is meant to ensure that future countervailing duties can be levied retroactively in case such policy is introduced. Where blends are imported, importers are obliged to specify the biodiesel content in the blends. The registration measure will remain in force for nine months. The bloc's biodiesel industry is confident that a decision on provisional duties will be taken in the next few months. **June** – Provisional anti-dumping duties have been imposed on imports of biodiesel and biodiesel blends from Argentina and Indonesia, which together account for 90% of the EU's imports. The measured results from an investigation launched last year in response to complaints by a group of EU biodiesel producers over sales of imported biodiesel below its normal value overseas. **August** – The environment committee of the EU Parliament has endorsed the EU Commission's draft proposal to cap the contribution of food-based biofuels in the transportation sector and to require reporting of estimated GHG emission levels for each biofuel, including mandatory accounting of ILUC effects. The committee recommended to limit the contribution of food-based (or "first-generation") biofuels to 5.5% of total energy consumption in 2020 – as opposed to the originally proposed 5%, and compared to the current level of about 4.5% – with a view to ensure that support granted to such biofuels does not unintentionally harm the environment or displace food production. The proposal represents a reversal from the EU directive approved in 2009, which fixed the contribution of renewable energy sources at 10%. The committee also proposed that "advanced" (or "second-generation") biofuels – that is, fuels made from agricultural residues, seaweed and certain waste products – account for at least 2% of total energy consumption and that their products could be promoted through special incentives. **September** – With a view to spur the development of "clean fuels" derived from nonfood sources (thus preventing unintentional harm to the environment or displacement of food production), in a plenary vote of the EU parliament the use of crop-based fuels for transportation has been capped at 6% by 2020. At the same time, second-generation biofuels from nonfood sources (like farm and industrial waste or lignocellulosic biomass) should account for at least 2.5%. **December** – Provisional anti-dumping duties on biodiesel imports from Argentina and Indonesia have been replaced with definitive ones. For the next five years, Argentina's producers will have to pay between 217 and 246 Euros per metric ton (corresponding to an ad-valorem rate of 24.6% on average), while Indonesian producers will be faced with duties ranging 122 to 149 Euros (or 18.9% on average). Both countries said they will formally appeal the duties.
	INDONESIA **February** – Reportedly, in an effort to further stimulate domestic consumption of palm oil-based biodiesel, the government raised the biodiesel distribution subsidy from Rp. 2,500 to 3,000 per liter, effective this year. The subsidy was introduced in 2009 when it amounted to Rp. 1,000 per liter. Furthermore, the government is expected to announce a new formula for determining biofuel prices, which would ensure that renewable fuel production remains profitable.

(*Continued*)

TABLE 2.6 (Cont.)

Year	Notable Policy Developments
	MALAYSIA **February** – With a view to check the rise in domestic palm oil inventories, the government plans to expedite implementation of the country's B5 program (i.e., mandatory blending of diesel fuel with 5% of palm oil-based biodiesel) at the national scale, while moving toward B10 by the end of 2013. Currently, the B5 blending requirement only applies to the three central states of Peninsular Malaysia, absorbing no more than 200,000 tons of palm oil per year. Nationwide B5 implementation is estimated to translate into an annual requirement of 500,000 tons of palm oil. **April** – After announcing the nationwide implementation of 5% mandatory fuel blending last January, the government is planning to move to B10 (10% mandatory blending) by mid-2014. Meanwhile, last month, B10 blends have already been introduced at the armed forces, the capital's city hall and at the Malaysian Palm Oil Board. In addition to pursuing environmental objectives, the policy aims at increasing palm oil usage in the domestic market and contributing to price stabilization. The same objectives are pursued by the recently announced creation of a partly state-owned biodiesel consortium. The consortium, which includes plantation and biodiesel companies as well as other investors, is expected to absorb one million out of the country's current 2.5 million palm oil stockpile. Reportedly, the consortium is poised to attract substantial government subsidies, the level of which will depend on the market price for palm oil. **August** – Launched in three central states of Peninsular Malaysia in 2011, mandatory commercialization of B5 (regular diesel blended with 5% palm oil methyl ester) is set to be extended to the entire country by July 2014, according to official sources. Once the program is implemented nationwide, biodiesel producers are expected to absorb 500,000 tons of palm oil per year, compared to 300,000 tons at present. Government officials expect that increased biodiesel demand will help absorb excess palm oil production and prevent the build-up of stocks, thus lending support to prices.
	CANADA **April** – Reportedly, the Canadian government plans to terminate biofuel production subsidies once its Ecoenergy for Biofuels programme (launched in 2008) expires in 2017. While biofuel production has been praised as a means for Canada to reduce its GHG emissions, official sources stated that the country's rapeseed and animal fat-based biodiesel industry has not been able to meet the national 2% blending target, creating the need to also import biodiesel. The petrol industry complained that domestically produced biodiesel often does not meet the specifications required for blending, forcing them to export biodiesel. While the government stopped accepting new applications for support in 2010, commitments for the 24 existing projects are going to be honored. Furthermore, the government is committed to preserve funds under its NextGen Biofuels programme. Industry sources informed that the termination of subsidies will make it impossible for domestic biodiesel producers to meet the national B2 mandate, which requires annual production of about 600 million tons.
	IRAN **April** – Reportedly, the country's first biodiesel production plant is ready to go online in Isfahan province. The plant has an installed capacity of 12,000 thousand tons per year and plans to use primarily jatropha oil as feedstock. Furthermore, the government plans to promote the use of food processing wastes as feedstock have been reported.

(Continued)

TABLE 2.6 (Cont.)

Year	Notable Policy Developments
	PHILIPPINES **August** – The National Biodiesel Board (a body composed of representatives from six ministries, the country's Coconut Authority and the Sugar Regulatory Administration) has approved an increase in the country's mandated biodiesel blend from currently 2% to 5% by end 2013 – two years ahead of the target set in the Biofuels Act of 2006. The planned increase would raise domestic demand for coconut methyl ester to about 350,000 tons per year, compared to 140,000 tons at present.
	CHINA **November** – Reportedly, China's biodiesel imports increased in recent months as trading companies started to take advantage of tariff and other trade incentives offered by the government to stimulate domestic biofuel consumption. While the importation of diesel is strictly regulated and local diesel sales are subject to consumption taxes, biodiesel imports from Association of Southeast Asian Nations (ASEAN) are duty-exempt (as long as they contain at least 30% biofuel). Although detailed import statistics are not available, Malaysia, Indonesia and Thailand are said to be the main import origins. **December** – Reportedly, the government will start taxing the consumption of fuels blended with biodiesel whenever the biodiesel content is less than 30%. The planned tax will match the existing consumption tax on conventional fuel. Blends with higher biodiesel content will remain tax exempt. Reportedly, the measures are meant to stop tax-free importation of low-rate blends into the country. Imports of such blends are said to have surged this year – encouraged by the absence of consumption taxes.
	SOUTH AFRICA **November** – The government informed that blending of diesel with biofuel – at a 5% rate – will be become mandatory from October 2015. By making blending obligatory and announcing the implementation date well in advance the government hopes to encourage investment into biodiesel development, production and infrastructure. Apparently, the economic incentives provided to date have not been sufficient to attract investors. The main feedstock for biodiesel production will be locally grown soybean, rape and sunflower seed.
2014	**INDONESIA** **February** – During 2014, Indonesia's biodiesel industry may absorb more palm oil than previously estimated. The government is determined to raise the share of locally produced biodiesel in domestic energy consumption, so as to scale down the nation's dependence on mineral oil and gas imports. In line with the recently introduced requirement that all diesel fuel used in the transportation and energy sector be blended with 10% biodiesel, in 2014, two state-owned firms (the nation's oil and gas company Pertamina and the country's electricity distributor PLN – Perusahaan Listrik Negara, an Indonesian state-owned company) expect to purchase a combined 5 million tons of crude palm oil – way above last year's level and a good 15% of total domestic palm oil output. If realized, these targets could temporarily curtail Indonesia's exports availabilities, potentially affecting world supplies. Reportedly, industry officials warned that higher domestic biodiesel consumption will require further improvements in the country's regulatory framework and pricing policy. **November** – Following sharp drops in domestic palm oil prices, the government decided to suspend the country's variable export tax on crude palm oil for the months of October and November. The measure aims to stimulate exports, thereby bringing down domestic

(*Continued*)

TABLE 2.6 (Cont.)

Year	Notable Policy Developments
	stock levels and halting the current decline in prices. The government's decision followed a similar move, in September, by rival exporter Malaysia. Prior to the suspension, Indonesia's exports were charged a 9% levy. The last time the tax was set at zero was in December 2009. One reason behind the recent surge in palm oil stocks has been lower than expected uptake by the local biodiesel industry. Apparently, the implementation of mandatory blending of diesel with palm-oil based biodiesel has fallen considerably behind schedule. With a view to raising domestic demand for palm oil, the government now plans to make B10 available nationwide by the end of this year, while the shift to B20 is envisaged for the year 2016.
	CHINA
	March – China has approved commercial use of bio-aviation fuel, in line with its policy to diversify the nation's fuel consumption and to reduce carbon emissions. The country's civil aviation authority licensed state-owned oil refiner Sinopec to use aviation fuel made from materials such as rapeseed, cottonseed and palm oil, as well as UCO. Reportedly, the measure makes China the fourth country in the world commercially using bio-jet fuel, after the US, France and Finland. Currently, China is the world's second largest aviation fuel consumer, and its demand is estimated to grow by 10% each year. According to observers, even if vast areas for growing oil-bearing plants as well as large amounts of waste oils seem to be available, high production costs are likely to prevent the large-scale application of vegetable oil-based jet fuel for the time being. Industry sources reported that it costs two to three times more to produce bio-based jet fuel than conventional fuel. The collection of UCO suitable for refining is said to be expensive, and it takes three tons of waste oil to generate one ton of bio-jet fuel.
	MALAYSIA
	March – The government confirmed that B5 blending with palm oil-based biodiesel (5% biodiesel and 95% petroleum-based diesel) will become mandatory nationwide in July 2014. The measure is expected to raise domestic demand for biodiesel from the current level of 200–300 thousand tons per year to 500 thousand tons.
	December – The government confirmed that the tax on crude palm oil exports will remain suspended until the end of the current year. The tax was temporarily lifted last September with a view to stimulate exports and help mitigate the downward trend in prices. Also, biodiesel policies continue to be used to stimulate domestic consumption and stabilize prices: in the transport sector in Peninsular Malaysia, the mandatory blending of diesel with 7% of palm oil methyl ester (B7) has commenced in November. Once implemented on a national scale and including the industrial sector, the B7 program is expected to utilize 700,000 tonnes of palm oil per year.
	CANADA
	May – To remain competitive with surrounding provinces, Ontario has introduced mandatory B2 blending (2% biodiesel in transportation fuel and heating distillate oil) at the provincial level, to be raised to B3 in 2016 and B4 in 2017. The new regulation also sets minimum GHG emission reduction requirements – initially at 30% but rising gradually to 70%. In Canada's four western provinces, local mandates at the B2 or B4 level are already in place. At the federal level, Canada introduced mandatory B2 blending back in 2011, allowing oil companies to blend renewable diesel anywhere in the country, provided the national average reached 2%. Under Ontario's new mandate, blending will take place within the province, potentially creating a market for locally grown soybean: an

(Continued)

TABLE 2.6 (Cont.)

Year	Notable Policy Developments
	estimated local biodiesel demand of 140,000 tonnes per year would open a market for 680,000 tonnes of soybean to be used as biodiesel feedstock.
	UNITED STATES
	May – The government announced that nearly $60 million will be made available to 195 producers to support the production of "advanced biofuel". The funding is provided through USDA's "Advanced Biofuel Payment Program" established under the 2008 Farm Bill and reauthorized in the 2014-Farm Bill.
	June – Minnesota, the first US state to have introduced mandatory blending (in 2002), decided to raise – during the summer months – its current B5 blending requirement to the B10 level. This year, the higher blend will be available at pumps from July through September, whereas next year, sales will start in April. While soy oil will remain the principal feedstock, at least 5% of Minnesota's biodiesel needs to be produced from nonagricultural resources. Reportedly, the temporary switch to B10 will translate into an additional demand of about 67,000 tonnes per year. In 2018, Minnesota could move to B20.
	December – $5.6 million in grants are going to be made available to 220 producers to support the production of "advanced biofuel". Payments will be made based on the amount of biofuel produced from renewable biomass other than maize kernel starch. Eligible feedstock includes vegetable oil, animal fat, crop residue and animal, food and yard waste. The funds will be provided through USDA's Advanced Biofuel Payment Program, which was established under the 2008-Farm Bill and reauthorized in the 2014-Farm Bill.
	ARGENTINA
	June – In an effort to support biodiesel producers, the government decided to temporarily suspend – until December 31, 2015 – a tax applying to domestic biodiesel consumption (amounting to 19% and 22% for fuel used for, respectively, transportation purposes and power generation), as well as to lower Argentina's export tax for biodiesel from 21% to 11%. In the previous year, domestic biodiesel production dropped sharply as a result of the antidumping tariffs introduced by the EU – the key export destination for biodiesel produced in the country. Thanks to the new tax concessions, industry officials expect this year's production and shipments to almost recover to the level of 2012.
	September – Starting on August 13, 2014, the tax collected on exports of soy oil-based biodiesel will be 12.5%, up from 11.2% in place since May 2014. Despite the small tax increase, Argentina should be able to sell biodiesel at attractive prices thanks to the current low price of soy oil, the principal feedstock. The tax was last revised in May 2014, when the government lowered export fees by almost half so as to support domestic biodiesel production and exportation.
	BRAZIL
	December – With a view to support the development of the country's relatively young biodiesel sector, the government decided to exempt the sale of vegetal-origin feedstock for biodiesel from paying selected taxes. The tax suspension will apply to both cooperatives producing feedstock and to processors and companies that store and market biodiesel feedstock.

(*Continued*)

TABLE 2.6 (Cont.)

Year	Notable Policy Developments
	INDIA **December** – The government has taken new steps to promote domestic production and consumption of biodiesel. According to the Railway Ministry, during the 2014–15 budget year, the country's fleet of locomotives will start using diesel blends containing up to 5% of biodiesel. Reportedly, the railway is India's single largest bulk consumer of diesel. Furthermore, the Road Transport Ministry is considering allowing biodiesel producers to sell their produce directly to domestic bulk consumers. By means of these two measures, the government hopes to stimulate the use of "clean" and domestically produced fuel while lowering the country's dependence on diesel.
2015	**EUROPEAN UNION** **February** – In France, the maximum level of biodiesel permitted in transportation fuel has been raised from 7 to 8%. The measure puts France's admixture rate above the EU-agreed limit of 7% – a development that has raised fears of fragmentation of the EU's single fuel market. French government officials presented the initiative as a technical adjustment to permit petrol companies to meet existing national targets. In France, the primary feedstock for biodiesel production is rapeseed oil of domestic origin. **May** – After consultations with the Council, the European Parliament agreed to limit, by the year 2020, the amount of crop-based biofuels (also called conventional or first-generation biofuels) that can be used in the transport sector to 7% – which compares to a 5%-limit originally proposed by the Commission and to 6% previously backed by the Parliament. Member states would also be required to set national targets for the share of so-called advanced biofuels, i.e., fuels based on waste and residues as well as on new sources of biomass such as seaweed. The draft law aims to accelerate the shift to alternative feedstock for biofuel production, thereby reducing GHG emissions caused by the growing use of farmland for biofuel crops.
	UNITED STATES **February** – The EPA has approved a policy change that could stimulate exports of biodiesel from Argentina to the US. The EPA accepted a request by the Argentine biodiesel industry to apply an alternative tracking method to prove that soy used to produce biodiesel was not grown on deforested land. By meeting this requirement, Argentine biodiesel qualifies to take part in the US' RFS program. EPA's decision attracted criticism from the US biodiesel industry, which claims that Argentina's new tracking system would be far less stringent than the current requirement and more difficult to verify. Market observers said that fresh biodiesel imports would intensify competition with local producers and pressure US biofuel prices.
	CHINA **March** – China's NEA published new policy guidelines for the development of the country's biodiesel industry. The guidelines' overall objective is to promote biodiesel use while protecting local resources. Waste and non-edible products (such as wood- and grass-biomass) will be favored over edible vegetable oils for use as feedstock. The new policy also focuses on (i) encouraging engine manufacturers to develop improved diesel motors for biodiesel usage, (ii) promoting corporate social responsibility, and (iii) improving recycling laws. Furthermore, biodiesel producers will enjoy tax advantages, and foreign investments into China's biodiesel industry will be encouraged. Interestingly, the government does not envisage to use mandatory blending requirements and

(*Continued*)

TABLE 2.6 (Cont.)

Year	Notable Policy Developments
	direct production subsidies to strengthen the biodiesel sector, and therefore, industry representatives raised questions about the implementation of the new guidelines at the local level.
	INDONESIA **March** -The government decided to change the method for setting the domestic retail price for biodiesel: starting in March 2015, pump prices will be determined based on the prevailing price of crude palm oil (plus processing costs and a 3% margin) rather than on the Singapore-based price for conventional diesel. Furthermore, biodiesel producers have been granted an almost threefold increase in state production subsidies, which passed from the IDR 1,400 ($0.11) per liter of biodiesel produced, paid until the end of 2014, to IDR 4,000 ($0.30) per liter. The annual volume of subsidized biodiesel has been set at 1.5 million tonnes, with government outlays to be financed from funds previously used to subsidize mineral fuel sales. The new method to set biofuel retail prices and the increase in producer subsidies are both aimed at stimulating domestic biodiesel production and guaranteeing its profitability amid falling international crude oil prices. Reportedly, the slump in crude oil quotations recorded since June 2014 has compromised the economic viability of the domestic biodiesel sector. **July** – The government has approved a regulation that requires exporters to pay a levy of $50 per tonne of crude palm oil and $10–30 per tonne of processed palm oil. The proceeds of the levy will primarily be used to fund the government's recently announced biodiesel subsidies. By promoting local biodiesel production (and thus the demand for palm oil), the measure is eventually expected to support domestic palm oil prices. Reportedly, the new levy, which should become effective in August, will only be due when the export price for crude palm oil falls below the $750 per tonne mark – the threshold at which the country's variable export tax kicks in.
	INDIA **April** – The central government confirmed that producers, suppliers and authorized dealers of biodiesel will be allowed to sell their product directly to all types of domestic consumers. Until now, biodiesel manufacturers were required to sell their products at a uniform price to designated OMC for distribution to end consumers. By liberalizing trade, the government hopes to foster the production and usage of biodiesel in the country. **November** – As an additional measure to encourage domestic biodiesel production and consumption, the government decided to exempt raw materials used for biodiesel production – i.e., oil and fat derivatives, methanol and sodium methoxide – from central excise duties. Biodiesel is already exempted from excise duties in its own right. Since 2014, the government has raised its efforts to promote domestic biodiesel consumption. In August 2014, it launched the sale of B5 in certain retail outlets of state-run OMC.
	THAILAND **May** – Seven percent mandatory blending transportation diesel with palm oil-based biodiesel has been reinstated in April – reversing a cut to 3.5% last January, when the government reduced the blending proportion in response to a temporary shortage in domestic cooking oil supplies. The move comes as domestic palm oil production and inventories have grown again, reaching burdensome levels and putting pressure on domestic prices.

(*Continued*)

TABLE 2.6 (Cont.)

Year	Notable Policy Developments
	SOUTH AFRICA **May** – The plans of a private-public joint venture to set up a large biodiesel plant have been put on hold. The plant was supposed to start production in 2017. Reportedly, delays in the issuance of a regulatory framework for the country's biofuel sector are behind the postponement. Regulations requiring a blended rate of 5 % biodiesel in regular transport diesel – as well as provisions for financial incentives to biofuel manufacturers – were expected to go into effect in October this year. Once implemented, the government's blending target should translate into an annual domestic requirement of 400,000 tonnes of biodiesel. The said investment project was supposed to match that demand, using about 1.1 million tonnes of rapeseed as feedstock.
	AUSTRALIA **July** – Originally planned to be phased in over the next five years, excise taxes for biodiesel will now be introduced over a longer period of 16 years, starting in 2016. The duty will be phased-in in equal increments until it reaches 50% of the excise rate applied to conventional diesel. The country's biodiesel industry welcomed the government's decision, saying that it affords investor certainty and enables producers to take a long term view, thus providing a sustainable footing for the sector's growth. Momentum for the use of biofuels as an alternative to mineral oils is reported to be growing in the country.
	PERU **August** – The country's consumer defense and competition authority initiated anti-dumping investigations about imports of pure biodiesel from Argentina. Reportedly, the move is in response to allegations by the domestic palm oil and biodiesel producers claiming unfair competition by Argentinian manufacturers. Back in 2009, Peru imposed antidumping and countervailing duties on biodiesel originating from the US. According to the private sector, although diesel fuel sold in Peru must contain 5% of biodiesel since 2011, efforts to promote domestic biodiesel production have been idled by the availability of competitively priced foreign biodiesel.
	BRAZIL **November** – Brazil's Council of Energy Policy (CNPE) authorized, effective January 1, 2016, the voluntary use – by certain consumer groups – of biodiesel blends exceeding the mandatory blending rate of 7%. Public and corporate transport fleets will be allowed to use biodiesel blends of up to 20% (B20), while 30% blends (B30) will be permitted in rail transport, agriculture and industrial uses. Furthermore, the ruling allows for experimental use of pure biodiesel (100%). Until specific assurances are provided by the car and machine manufacturers, the higher blends will not be available at public pumps. Government plans to increase mandatory blending from 7% to 10% in 2020 remain in place.
	ARGENTINA **December** – Effective October 2015, Argentina's biodiesel export tax was lowered to 3.3%, down from the 9.8% in place since May 2015. The new rate is the lowest since the government introduced its variable tax system in 2012. The duty has been reduced to support the country's export-oriented biodiesel sector, which is struggling to survive, amid depressed fossil oil prices and the imposition by the EU, of antidumping tariffs.

(*Continued*)

TABLE 2.6 (Cont.)

Year	Notable Policy Developments
	With few alternative markets, Argentina's biodiesel production has fallen sharply, resulting in a capacity utilization rate of only 40%. In addition to lowering taxes, the government also reduced the domestic biodiesel reference prices by one percent, with a view to stimulating local biodiesel consumption.
	PHILIPPINES **December** – The city of Davao, Philippines, together with public and private entities from Japan, launched a project to collect UCO from households and businesses for biodiesel production. Reportedly, although UCO is classified as toxic and hazardous, it is currently sold on the city's black market.
2016	**AUSTRALIA** February – The government of Queensland, the country's third most populous state, ruled that biodiesel and renewable biodiesel have to make up 0.5% of local diesel sales as of mid-2017 – following the example of New South Wales, where diesel sales contain 2% of biodiesel since 2010. At the same time, the federal government is continuing its gradual phasing-out of support provided to biofuel producers in the form of subsidies and tax concessions.
	UNITED STATES **February** – The US Congress passed the extension of the $1 per gallon ($0.26 per liter) tax credit for biodiesel blenders which had expired at the end of 2014. The tax credit applies retroactively to January 1, 2015, and will remain in place until December 31, 2016. A bid to reform the subsidy – narrowing the scope of the credit to domestic production – has been turned down by policymakers. Since blenders utilize both domestic and imported product (depending on prices), the modification would have made imported biodiesel ineligible for the credit and thus less competitive. Conversely, the subsidy's extension in its current form could – given the availability of competitively priced biodiesel on the world market – lead to a further increase in US biodiesel imports in 2016, at the expense of domestically-produced biodiesel. **June** – The US EPA proposed a mandatory volume requirement for biomass-based diesel in 2018 of 2.1 billion gallons – compared with final standards of 1.9 billion gallons and 2.0 billion gallons in 2016 and 2017, respectively. The country's current production capacity is estimated at 2.1 billion gallons. In Iowa, the state with the second largest biodiesel production capacity in the US, the local government decided to extend its biodiesel production credit – originally set to expire at the end of 2017 – through 2024. The producers' credit will remain at 2.0 US cents per gallon for the first 25 million gallons of production per plant. In addition, the state's retail tax credit is set to be extended and expanded. While retailers will continue to receive a tax credit of 4.5 US cents per gallon for sales of diesel containing at least 5% biodiesel (B5), from 2018 through 2024, the B5 incentive will drop to 3.5 US cents per gallon, but an additional incentive of 5.5 US cents per gallon will take effect for diesel containing at least 11% biodiesel. **November** – A US district court has ruled that biodiesel blending mandates applied in the state of Minnesota since 2005 do not conflict with the federal RFS, and that the RFS, therefore, does not preempt Minnesota's mandates. Under Minnesota state law, diesel fuel sold within the state is required to contain a specific percentage of biodiesel (currently 10%, with a transition to 20% scheduled for 2018 – see MPPU 2018). The

(*Continued*)

TABLE 2.6 (Cont.)

Year	Notable Policy Developments
	plaintiffs – oil, gas, trucking, auto manufacturer and auto dealership trade associations – had argued that Minnesota's per gallon blending requirements, geographical blending obligations and timing restrictions are not compatible with the national RFS. Minnesota was the first state to pass a biodiesel blending law, but other US states have subsequently followed suit.
	December – The US EPA increased the targets for renewable fuel consumption under the country's RFS. Although the standards have been kept below the statutory targets set in 2007 to reflect changing market realities, the steadily increasing volumes continue to support Congress's intent to expand total consumption of renewable fuels, EPA said. The agency finalized the volume requirements and associated percentage standards for total, conventional and advanced renewable fuels for the year 2017, as well as the volume for biomass-based diesel in 2018. The volume of total renewable fuel will rise to 19.28 billion gallons in 2017, up 6.5% form 2016 but significantly below the target originally set by Congress. Within that total, the volume of nonadvanced/conventional biofuels (mostly maize-based ethanol) will increase to 15 billion gallons, finally meeting the congressional goal; and advanced biofuels, which are required to achieve at least 50% lifecycle GHG emission reductions, are set at 4.28 billion gallons – up 19% from 2016, but still below the original congressional target. Under the advanced biofuels category, biomass-based diesel (produced from vegetable oils, notably soybean oil, and animal fats) is set to grow by 100 million gallons in both 2017 and 2018, that is, to volumes twice the minimum target mandated by Congress. In addition, vegetable oil-based diesel may also qualify under the "undifferentiated advanced biofuels" category. Given that several of these final targets exceed the proposals unveiled by EPA earlier this year, a higher than previously anticipated volume of feedstock, in particular of soybean oil, might be required to fulfill the new US mandates – a prospect that triggered price spikes across vegetable oil markets worldwide.
	ARGENTINA
	February – In February 2016, the export duty on soy oil-based biodiesel was raised from 1.62% – the lowest rate since the introduction of the variable tax regime in 2012 – to 3.89%.
	March – In March 2016, Argentina's export duty on biodiesel was raised to 6.4%. This follows an increase in February to 3.89% from a historical low of 1.62% in January. The mandatory biodiesel blending rate for transport diesel remains unchanged at 10%, the level in place since January 2014. It seems that Argentina tries to maintain a large export of biodiesel under the anti-dumping tariffs imposed by the EU.
	September – Argentine biodiesel exports have been taxed at, respectively, 3.96 (May), 5.04 (June), 7.15 (July) and 4.99% (August). The circumstance that, since mid-2015, the Government modified the tax on a monthly basis has prompted the local biodiesel industry to call for less frequent changes so as not to disrupt its business.
	BRAZIL
	April – Legislation lifting the mandated share of biodiesel in diesel transportation fuel from the current 7% to 10% has been passed in Brazil. The mandatory blend rate will first be raised to 8% in April 2017, and subsequently to 9% in 2018 and 10% in 2019. The new legislation also requires the National Council for Energy Policies to test the feasibility of a 15% blend over the coming 36 months. The measures are aimed at boosting domestic biodiesel production, reducing petroleum import dependency and lowering pollutant emissions. Based on Brazil's fuel consumption, raising mandatory blending by one

(*Continued*)

TABLE 2.6 (Cont.)

Year	Notable Policy Developments
	percentage point lifts annual domestic biodiesel usage by about 0.5 million tonnes. The government's biodiesel policy also promotes family farming, as Brazil's "Social Fuel Seal" program requires a certain amount of raw materials to be sourced from family farms. In Brazil, soybean is expected to remain the main feedstock used for biodiesel production.
	THAILAND **June** – In Thailand, where the mandatory blending of diesel transportation fuel with 7% of palm-oil based biodiesel is in place since 2014, the voluntary nationwide sale of B10 (diesel blends containing 10% biodiesel) is planned for 2018. Policymakers are working on industry standards for the higher blends and may introduce tax incentives to make B10 more attractive. B7 will remain commercially available alongside B10. For vehicles run by state agencies and the military, B10 usage will be mandatory. **July** – The government has signed MoUs with private energy and logistics companies to use B20 biodiesel for their heavy vehicles. THB 115 million ($3.3 million) from the country's Energy Conservation Promotion Fund will be made available to subsidize purchases of the alternative fuel by the signatory companies. **September** – The government decided to reduce the mandatory blending of transportation diesel with palm oil-based biodiesel from 7 to 5% in July 2016, and further to 3% in August 2016. Blending requirements have been lowered in a bid to end a temporary shortage in domestic cooking oil supplies. Reportedly, national palm oil supplies have been lower than earlier anticipated (because of poor yields linked to El Niño), raising concern that soaring cooking oil prices could harm consumers. No date has been provided for the restoration of the 7% blending rate.
	NEPAL **July** – The Government of Nepal formed a committee to examine the possibility of producing biodiesel from jatropha seed. Over 500,000 hectares of unused land are deemed suitable for jatropha cultivation, and preliminary studies claim that several regions have a high potential for commercial jatropha farming. The government is pursuing biodiesel initiatives with a view to reduce the country's dependence on imported fossil fuels, while reducing carbon emissions. Reportedly, the country still has to put in place the necessary infrastructure for biodiesel production and marketing.
	PHILIPPINES **May** – The country's Department of Energy issued new specifications for coconut methyl ester (coconut oil-based biodiesel). Allegedly, the revised national standard (which features higher iodine and oxidation stability values, a lower maximum sulfur content, and a cold soak filterability test) raises the bar for quality beyond the globally accepted American Society for Testing and Materials (ASTM) biodiesel standard. Compliance with the new standard has become mandatory for all biodiesel produced and sold in the country.
	EUROPEAN UNION **August** – European Commission outlines measures to accelerate the EU's transition to a low-carbon economy, the Commission proposed to gradually phase out – from 2020 – conventional food-based biofuels, while providing incentives for the development of more advanced ("second-generation") biofuels, including fuels produced from agricultural residues. According to the document, conventional biofuels play a limited role in decarbonizing the transport sector and should not receive public support after 2020. The

(*Continued*)

TABLE 2.6 (Cont.)

Year	Notable Policy Developments
	Commission also proposed – for the first time – a set of ambitious, binding GHG emission reduction targets for the transport, buildings, agriculture, waste, land-use and forestry sectors – i.e., the sectors that are not included in the bloc's Emissions Trading System. The proposed targets are country-specific and refer to the period 2021–2030, using 2005 emission levels as a baseline.
	European Court of Auditors pointed out that weaknesses in the system certifying sustainable biofuels could undermine the basis of the EU's 2020 targets for renewable energy in transport. Under EU rules, member states can only use biofuels certified as sustainable to reach their targets. While most biofuels are certified through voluntary schemes recognized by the European Commission, the auditors expressed the opinion that, because of weaknesses in the Commission's recognition procedure and in the supervision of the schemes, the EU's biofuel certification is not fully reliable. They also pointed out that member country statistics might overestimate the use of sustainable biofuels by including fuels whose sustainability was not fully verified. ECA called on the Commission to ensure that all certification schemes (i) assess whether biofuel production entails significant socio-economic risks and indirect land-use change; (ii) verify that feedstock producers comply with environmental requirements for agriculture; and (iii) provide sufficient evidence of the origin of waste and residues used for biofuels.
	November – On 15 September, in a parallel domestic proceeding based on cases filed by individual Argentine and Indonesian biodiesel producers, the EU's General Court ruled that the antidumping duties imposed on biodiesel imports from Argentina and Indonesia infringe basic regulations and should be annulled. Similar to the WTO case, the EU judges found that the European Commission's calculations of the costs of biodiesel production in Argentina and Indonesia were based on incorrect considerations. The European Commission and the EU's biodiesel industry have two months to file an appeal to the European Court of Justice.
	INDONESIA
	August – State-owned power company PLN informed that it cannot comply with government rules on burning diesel containing 30% of palm oil-based biodiesel (known as B30) because such blends would damage its generators. Under regulations released last year, starting in January 2016, Indonesian power firms are required to use B30. Commenting on PLN's statement, market observers pointed out that lower than expected uptake of biodiesel by the country's electricity sector could weigh on domestic and international palm oil prices – given that domestic supplies (and hence export availabilities) of palm oil could experience a significant rise. The government's 2016 target for total domestic uptake of palm oil-based biodiesel (by both the transportation and power generation sectors) is 6.1 million tonnes, a level industry experts consider to be out-of-reach.
	MALAYSIA
	December – Contrary to earlier announcements, the government decided to defer the planned shift to B10 for the transport sector and to B7 for the industrial sector – i.e., diesel fuel containing, respectively, 10% and 7% of palm oil-based biodiesel – to a later, unspecified date. Government officials explained that the prevailing large price gap between mineral oil and palm oil has significantly raised the cost of implementing the higher mandates, risking to unduly burden consumers. The shift was deferred until a more suitable time, that is, until the price gap diminishes. The country's palm oil producers remain in favor of the move, which, by raising annual domestic palm oil use by an

(*Continued*)

TABLE 2.6 (Cont.)

Year	Notable Policy Developments
	estimated 800,000 tonnes, would help bring down the country's palm oil stocks and support prices.
	NEW ZEALAND **August** – The country's first commercial-scale biodiesel plant was expected to commence production in August 2016. The concerned company is using inedible tallow (a by-product of the meat industry typically used in the manufacturing of soap and candles) as feedstock, likely absorbing about 12% of the country's inedible tallow output. The plant's annual biodiesel output is pegged at 17,000 tonnes.
	UNITED KINGDOM **September** – According to industry sources, palm oil is no longer present in the renewable biodiesel that is blended into fossil fuels in the UK. Reportedly, palm oil content started dropping in 2012, and UK biodiesel – whether produced in the UK or imported – is now largely made from the waste feedstock, in particular, UCO.
	PERU **November** – The Peruvian Government has announced the replacement of provisional anti-dumping duties imposed last February on biodiesel imports from Argentina with definite ones. The duties, which are set to remain in place for five years, will be charged at variable rates (ranging from $122 to $192 per tonne), depending on the biodiesel manufacturer in Argentina. The measure is meant to shield biodiesel production in Peru from allegedly unfair competition by Argentine producers. Argentina is the main supplier of biodiesel to Peru, where consumption of B5 blends (i.e., transportation diesel containing 5% of biodiesel) became mandatory in 2011.
2017	**UNITED STATES** **February** – Renewing national efforts to promote research on bio-based fuels, the US Department of Energy has allocated funds to a research project on camelina sativa, an edible oil crop adapted to northern climates and suitable for cultivation as a rotation crop in dry areas and on thin soils. Camelina will be studied for its potential as a biofuel feedstock. In recent years, numerous tests have been conducted in the US and elsewhere with camelina-based transport and aviation fuels, and camelina oil has gained approval as biofuel feedstock by the country's EPA. **April** – Claiming that Argentina and Indonesia are dumping biodiesel onto the US market, the US biodiesel industry filed an antidumping and countervailing duty petition with the US Government. According to the petition, Argentine and Indonesian producers are selling biodiesel into the US market at prices substantially below their costs of production. The petition also alleges that Argentine and Indonesian producers enjoy trade and market-distorting subsidies. US biodiesel imports from the two countries are said to have surged in recent years, taking market share from US manufacturers. Reportedly, the Indonesian Government has submitted a complaint to the WTO over the US industry's petition, while Argentine state officials have rejected the US claims as unfounded. **June** – The Iowa Infrastructure Fund Bill, which provides $3 million in funding for the state's Renewable Fuels Infrastructure program, has been extended for an additional year until June 30, 2018. Under the program pump operators receive assistance for the

(*Continued*)

TABLE 2.6 (Cont.)

Year	Notable Policy Developments
	conversion of their equipment to raise renewable fuel sales. Iowa is the US' leading producer of biodiesel. **June** – (1) Legislation concerning the renewal of the $1.00-per gallon biodiesel tax credit – which expired on December 31, 2016 – has been introduced in the US Congress. The bill envisages the subsidy's extension for 3 years – retroactive from January 1, 2017, and through December 31, 2019. However, the main change concerns the stage in the supply chain that will receive the tax incentive: under the proposed bill, the subsidy would be provided to domestic biodiesel producers rather than blenders. Since its introduction in 2005, the tax incentive was directed to blenders, making no distinction whether the feedstock used was produced domestically or imported from abroad – a provision that has stimulated the importation of competitively priced biodiesel into the country, in part displacing US product. The proposed shift to a producers' credit is aimed at preventing foreign manufacturers from accessing the tax benefit. However, the reform is opposed by some, who argue that the shift would increase profits for a limited number of producers, possibly reducing the overall availability of biodiesel and driving up its price. Previous bids to convert the blenders' credit into a producer incentive have been turned down by the US Congress. (2) A separately introduced bill proposes to put an end to federal tax credits for biodiesel produced with animal fats. Allegedly, the current policy distorts the domestic market for animal fats by diverting this important raw material away from the manufacturing of cleaning and personal care products. **July** – Five percent biodiesel blends in home heating oil will become mandatory in New York City and downstate New York counties by, respectively, October 1, 2017 and July 1, 2018. With a view to reduce the region's carbon footprint, New York City mandated 2% blends back in 2012, with subsequent increases set to lead to 20% blends by 2034. Reportedly, the city is the largest municipal consumer of heating oil in the US. **August** – The EPA published its proposals for mandatory consumption of total, conventional and advanced renewable fuels in 2018, as well as biomass-based diesel in 2019. As in past years, several of the proposed volumes range below the statutory targets set by Congress in 2007, allegedly reflecting the need to account for changing market realities. For 2018, EPA proposed (i) a slight year-on-year drop in the volume of total renewable fuels, which would keep volumes significantly below the congressional target; (ii) to leave nonadvanced/"conventional" biofuels (mostly maize-based ethanol) unchanged compared to 2017, compliant with the original targets; and (iii) to lower advanced biofuels from the 2017 level, hence falling short of statutory volumes. Under the "advanced biofuels" category, volumes of biomass-based diesel produced from vegetable oils and animal fats (which were raised in past years) would remain unchanged in 2019. Furthermore, "biomass-based diesel" would continue to qualify as well under the "undifferentiated advanced biofuels" category. **August** – Directly related to the proposal above, in July 2017, a US court ruled that the methodology used by EPA to justify past reductions in biofuel consumption mandates (relative to the statutory targets set by Congress) was incorrect. The ruling was made in response to a petition filed in 2016 by an alliance of biofuel advocacy groups. According to the court ruling, rather than taking into account renewable fuel supplies available to refiners and importers, EPA primarily considered biofuel demand expressed by consumers – a consideration not allowed under the relevant law. The ruling could force EPA to revise upward the consumption targets set for 2016. Reportedly, the court decision

(*Continued*)

TABLE 2.6 (Cont.)

Year	Notable Policy Developments
	resulted in US renewable fuel credits climbing to multi-month highs, while share prices of mineral oil refiners dropped. **August** – The US biodiesel industry filed an additional allegation with the US Department of Commerce (DOC) claiming that "critical circumstances" exist with respect to imports of low-cost biodiesel from Argentina. Based on an earlier petition, in April 2017, DOC launched preliminary investigations into the named imports for possible dumping and subsidization, with preliminary determinations due later this year. In principle, the "critical circumstances" provision allows for the imposition of countervailing measures prior to preliminary determinations, possibly offering relief to affected parties in the form of retroactive duties – protection aimed at deterring exporters from boosting shipments before countervailing measures are introduced. Claiming that shipments of Argentine biodiesel into the US surged since the industry filed its first petition, US biodiesel producers have taken the additional step of invoking the "critical circumstances" clause. **October** – The US EPA is considering reducing the mandatory renewable fuel consumption targets for 2018 and 2019 that it proposed earlier this year and is seeking public comments in this regard. Reportedly, EPA decided to revisit its original proposal as new data on production, imports and costs of biodiesel have become available. Allegedly, the price of biodiesel to blenders as well as the price of biodiesel blends to consumers has increased following the expiration of the biodiesel tax credit in December 2016. Moreover, prices may continue to rise as a result of the recent preliminary determination of countervailing duties on imports of biodiesel from Argentina and Indonesia, said EPA. Therefore, the agency is concerned that the originally proposed targets may lead to inadequate domestic supplies of biofuel to consumers and is hence evaluating the possibility of setting lower targets. The domestic biodiesel industry questioned EPA's rationale, stressing that the country's biodiesel production capacity was sufficient to meet the original targets. Recently, the American Soybean Association even urged the agency to raise the target for biomass-based diesel, arguing that biodiesel production creates a value-added market for the country's abundant soybean oil supplies and that the country's growing soybean production could support higher demand for biodiesel without leading to price increases. **October** – the US Commerce Department set – in addition to the preliminary countervailing duties imposed earlier this year – preliminary antidumping duties on imports of biodiesel from Argentina and Indonesia, claiming that Argentine and Indonesian biodiesel was sold in the US at dumping margins of, respectively, 54–70% and 51%. The Department is expected to announce its final antidumping decision in January 2018. **December** – Following consultations with lawmakers, the US EPA decided to abandon its recent proposal to lower the mandatory targets for renewable fuel production in 2018 and 2019. The definitive targets announced for 2018 entail (i) a fractional increase – compared to 2017 – in the total renewable fuels volume; (ii) unchanged levels of nonadvanced/conventional biofuels; and (iii) a marginal rise for advanced biofuels. Under the "advanced biofuels" category, volumes of biomass-based diesel have been set at 2.1 billion gallons for both 2018 and 2019 – which compares to 1.9 billion gallons in 2017. The country's biodiesel industry criticized the final targets, arguing that these remain well below the sector's production capacity.

(*Continued*)

TABLE 2.6 (Cont.)

Year	Notable Policy Developments
	RWANDA **April** – The Rwanda government decided to abandon a $35 million biodiesel project launched four years ago. Reportedly, the government realized that the pilot project was not viable due to insufficient availability of feedstock and high costs of production. The project feasibility study had identified nonedible jatropha oil as key feedstock, but eventually, the plant was discovered to be unsuitable for the country's climate. Concerns over using scarce arable land for uses other than food production were also cited.
	SINGAPORE **June** – Air carrier Singapore Airlines, in partnership with Singapore's Civil Aviation Authority, started operating a series of biofuel-powered intercontinental flights. The flights will be powered by a blend of conventional jet fuel, hydro-processed esters and fatty acids produced from UCO. The project is geared toward reducing the carrier's carbon emissions.
	ITALY **June** – Italy's National Consortium for the Collection and Treatment of Used Oils and Fats (CONOE) signed a Memorandum of Understanding to supply used vegetable oils and fats to multinational oil/gas company ENI. The company, which has experience in producing renewable diesel, jet fuel and liquefied petroleum gas from palm oil, plans to process approximately 1 million tonnes of used oils/fats annually, starting in 2018. Reportedly, the new partnership also includes joint activities to promote the collection of used oil/fat at the household level, where the product habitually goes to waste. ENI expects to achieve significant reductions in carbon emissions and freshwater consumption by switching to used oils/fats.
	ARGENTINA **July** – The Argentine government suspended the country's export duty on biodiesel for the months of June and July. In April and May, the variable duty, which is reviewed on a monthly basis, stood at, respectively 7.05% and 0.13%. The duty's suspension is expected to improve the competitiveness of the Argentine biodiesel on the world market, at a time when the country's export prospects are threatened by reduced demand from key buyers, notably the US. **September** – Regarding the EU's commitment to bring the bloc's antidumping duties on biodiesel imports from Argentina into conformity with a recent WTO ruling, the two parties, Argentina and EU, agreed to extend the EU's time limit for removing or altering the named tariffs until September 28, 2017 (as opposed to the originally agreed upon date of August 10, 2017). **September** – The US DOC announced affirmative preliminary determinations in the countervailing duty investigations related to Indonesian and Argentine exports of biodiesel to the US. DOC claimed that Argentine and Indonesian exporters received subsidies of, respectively, 50–64% and 41–68%. The department also determined that "critical circumstances" exist in both cases, which allows for the retroactive collection of duties. Final duty determinations are due on November 7, 2017. While DOC finalizes its investigations, biodiesel importers will be required to pay cash deposits on purchases from Argentina and Indonesia based on the indicated subsidization rates, going back to May 2017.

(*Continued*)

TABLE 2.6 (Cont.)

Year	Notable Policy Developments
	September – Minnesota, the first state in the nation to mandate biodiesel use back in 2005 – is set to raise its biodiesel standard from 10 to 20% in May 2018. Accordingly, only diesel containing 20% biodiesel may be sold between May and October, when fuel stations will switch back to 5%, as biodiesel turns from fluid to gel under colder temperatures. The higher blending mandate is meant to (i) help preserve and protect Minnesota's air and water quality, (i) reduce the state's reliance on fossil fuels, and (iii) support demand for locally grown biodiesel feedstock, notable soybean. Fuel producers and the trucking industry expressed concern about the new measure, stating that the infrastructure required to handle the higher blends was not yet available.
	January 2018 – The Government announced a rise in the country's variable export tax on biodiesel from zero percent applied in December 2017 to a fixed rate of 8%, effective January 1, 2018. Considering that, at the same time, the January export tariff rate for soy oil is going to be lowered to 26.5% (from 27% applied in December), the tax difference between the two products is set to drop from 27% to 18.5%. Reportedly, the reduced tax differential could result in negotiations with the US regarding last year's introduction of antidumping and countervailing duties on imports of Argentine biodiesel (N.B. The US likened Argentina's tax differential to a subsidy for the country's biodiesel producers). Market observers pointed out that the higher taxation of biodiesel exports could negatively affect Argentina's sales of the fuel to the EU.
	JAMAICA
	July – The state-owned Petroleum Corporation of Jamaica has developed diesel containing 5% of castor oil-based methyl ester and is testing the biodiesel on motor vehicles. Reportedly, the Jamaican government is envisaging the introduction of transport biodiesel, with a view to cut the country's energy import bill and contribute to the reduction of GHG emissions.
	NORWAY
	July – A resolution passed by the Norwegian Parliament calls on the country's government to take steps to end – on environmental grounds – public procurement of palm oil-based biodiesel, while promoting the use of sustainably produced, advanced biofuels.
	CHINA
	July – A subsidiary of China's state-owned oil company Sinopec is planning to set up a UCO biofuel plant in Eastern China. The plant would convert 100,000 tonnes of UCO into 30,000 tonnes of aviation-grade biofuel a year. Reportedly, the fuel would be sold to airlines operating long-haul international flights, especially to countries that charge high emission taxes.
	EUROPEAN UNION
	August – In line with pledges to bring the bloc's antidumping duties on biodiesel imports from Argentina and Indonesia into conformity with WTO rules, the EU Commission has tabled a proposal for lowering the duties that are in place since November 2013. Under the proposal, the duties on imports from Argentina would be reduced from around 25% to 9%, and those on Indonesian produce from about 19% to 5%, according to media reports. The EU's agriculture and biodiesel lobbies strongly criticized the proposal, arguing that their sectors would be severely impacted by the reduction in duties, which, allegedly, would trigger an influx of low-cost biodiesel from Argentina and Indonesia

(*Continued*)

TABLE 2.6 (Cont.)

Year	Notable Policy Developments
	into the bloc. The Commission decided to postpone the final vote on the matter to September 2017. **October** – Following consultations with the Member States, the European Commission decided to reduce the bloc's custom duties on biodiesel imports from Argentina, effective September 19, 2017. To conform with a recent WTO ruling, the EU lowered its anti-dumping duties to between 4.5% and 8.1% – as against the 22–25.7% range in place since May 2013. The EU's decision has come at a time when the US chose to implement its own measures to restrict biodiesel imports from Argentina. Accordingly, Argentina's export-oriented biodiesel industry welcomed the possibility to resume shipments to the EU market. **January 2018** – The European Council outlined its position regarding the reform of the bloc's RED for the period 2021–2030. The Council proposed to retain the existing target of 27% renewable energy in total energy consumption in 2030. In the transport sector, the 14% target for each member state would be maintained for 2030, with a sub-target of 3% for "advanced biofuels". Also, the existing 7-percent cap for first-generation biofuels (such as vegetable oil-based biodiesel) would be maintained. If a member state sets a lower cap, it would have the option of lowering its overall target for renewables in transport. By comparison, the proposals tabled by the European Commission envisaged a progressive reduction of food-based fuels and their replacement with second-generation biofuels. Also, the European Parliament called for more ambitious overall targets and recommended to distinguish between first-generation biofuels produced using sustainable practices and those purportedly produced in an unsustainable manner. The European Commission confirmed that rapeseed grown in Canada and Australia would remain eligible for entering the EU market as environmentally-friendly feedstock for biodiesel, as the crops have been shown to meet the bloc's increasingly stringent requirements (From January 2018, all biodiesel feedstock need to deliver GHG lifecycle savings of 50–60% compared to fossil fuels – up from 35% in 2017). For Australia, the EU represents a top rapeseed export market, with the bulk of deliveries absorbed by the bloc's biodiesel industry.
	THAILAND **May** – The mandatory biodiesel blending rate of 7% has been reintroduced in May 2017. The rate had been temporarily lowered to 5 and 3% last year following shortages in domestic palm oil supplies and consumer price hikes. By contrast, in recent months, producer prices dropped sharply amid rising palm oil output, triggering Government measures to support domestic palm oil uptake, including the recent revision in the biodiesel blending rate.
	UNITED ARAB EMIRATES **October** – The Emirates National Oil Company (ENOC) announced the launch of a biodiesel blend destined for the country's commercial and industrial segment. The fuel, which contains 5% biodiesel produced from vegetable oil, UCO or animal fat, is expected to help reduce the country's GHG emissions. The measure is in line with the government's Energy Plan leading up to the year 2050, which calls for an energy mix combining renewable, nuclear and clean energy sources. Reportedly, the new blend is suitable for use in new and existing trucks as well as heavy construction equipment, without need to upgrade engines and fuel storage facilities.

(Source: Food and Agriculture Organization or FAO, 2018) [14]

2.3 Biofuel Generations, Pros and Cons

Due to the concern on the availability of fossil fuels as well as the environmental impacts from the uncontrolled consumption of carbon-based fuels in the future, considerable attention has been paid to renewable resources. However, production and industrialization of renewable sources could be more expensive than those of nonrenewables, and they may run into some difficulties when executing them.

Three types of biofuels have been generated named as first-, second-, and third-generation biofuels. Whereas, they are normally characterized by their sources, limitations, and technologies. For example, the main drawback of the first-generation biofuels is that they are produced from food sources (e.g., biomass – sugar, starch, or vegetable oil). Nonfood biomass, such as corn stover (leaves, stalk, and stem of corn) and cellulosic sources that grow alongside food crops are categorized as the second-generation biofuels. But the second generation still competes with food production for land use. In addition, it is proved that their cultivation would consume too many nutrients from the soil, which should be replenished by fertilizers. Additionally, pretreating biomass to release the trapped sugars from the second-generation biofuel is energy and materials consuming. Thus, they are replaced by the third-generation biofuels, such as algae.

The third-generation biofuel is the best alternative fuel thus far in term of low-cost, high-energy contents, and completely renewable sources of energy. However, there are still some challenges in making them commercially and economically feasible. For example, algae biodiesel is produced from a nonfood biomass; thus, it may have less social impacts than other types of feedstock. Algae can grow in areas unsuitable for other generation crops in terms of water and arable land used. Additionally, it can be grown using sewage, wastewater, and saltwater, such as oceans or salt lakes. However, further research is necessary to better understand the extraction process in order to make it financially competitive to other fossil fuels. Unlikely, oil price fluctuations from 2014 to 2016 and the absence of a coherent and clear biofuel policy deter the investors away from this industry (Hochman, 2014) [15].

As discussed earlier in this chapter and in other chapters, unlike other renewable energy sources, biomass can be directly converted to liquid fuels, that is, biofuels, for the practical purposes, for example, for the transportation needs. Biodiesel and bioethanol are the most common types of biofuels. However, when biodiesel is produced from vegetable oils and animal fats, there are concerns that the feedstock may compete with food supply in the long-term. Further, feedstock costs account for 60% to 75% of the total production cost of biodiesel fuel (Hochman, 2014) [15].

Because it is widely recognized that biodiesel can reduce emission levels of some pollutants, and furthermore, usage of biodiesel will allow a well-balanced socioeconomic structure to be sought between agriculture, social, economic, political, and environmental developments in the future (Demırbas, 2017) [17], biodiesel feedstock diverted from the food supplies is imperative to make biodiesel more accessible. Hence, the recent studies on biodiesel feedstock are focused on nontraditional sources that produce nonedible oils as a feedstock for biodiesel production. Animal waste, aquatic plants and algal oil, as well as organic municipal and industrial waste are promising sources for biodiesel production. In this section, nonconventional sources, such as spent coffee ground (SCG), Soapnut seed oil (SSO), Jatropha oil, WCO, and algal oil are discussed from various point of views.

SCGs or coffee residue wastes provide excellent raw material for biofuel production. Oil from SCGs composed mainly of 40.2% linoleic acid, 35.9% palmitic acid, 10.7% oleic acid, and 7.5% stearic acid. The oil presents high amounts of palmitic and linoleic acids, which has a potential application for biodiesel production (Rocha et al., 2014) [18]. The SSO is found to mainly have 9.1% free fatty acid, 84.4% triglycerides, 4.9% sterol and 1.6% others. Compared to Jatropha oil, which contains 30–40% oil content (Ambat et al., 2018) [19] consisted of approximately 14% free fatty acids, SSO has approximately 5% less fatty acids than Jatropha oil (Chhetri et al., 2008) [20]. Oleic acid was reported to be a dominant fatty acid in both SSO and Jatropha biodiesel (Chhetri et al., 2008) [20].

WCO is composed of unsaturated fatty acids with 52.9% of mono-unsaturated and 13.5% of di-unsaturated fatty acid (Omidvarborna, et al., 2015) [21]. The composition of fatty acids (in wt%) in WCO mainly contains linoleic and oleic fatty acids (18 carbon atoms). However, according to the nature of the source oils, the composition of produced biodiesel may vary. The common fatty acids in algae oil are composed of saturated and unsaturated fatty acids with 12–22 carbon atoms (Mata et al., 2010) [22]. Although current efforts and investment are obtaining much attention toward producing algal biodiesel in industrial scale, cost per unit area and optimum nurturing environment should be monitored and controlled through large-scale production. Such a large-scale industrial process may become economical when it is combined with sequestration of CO_2 from flue gas emissions, with wastewater remediation processes, and/or with the extraction of high-value compounds for making value-added products in other process industries.

From the economical point of view, animal fats ($0.4 – $0.5/L), vegetable oil ($0.54–$0.62/L) and waste grease ($0.34–$0.42/L) can be counted as the cheapest raw materials for biofuel production. However, considering the cost of traditional transesterification of vegetable oils ($0.6–$0.8/L) (IEA, 2007) [23], the production prices for biodiesel become approximately $1.0 up to $1.5/L. Currently, pre-tax petroleum diesel in the US is about $0.18/L and in some European countries ($0.20–0.24/l), and therefore, it is very hard for biodiesel to compete with the petroleum diesel in the fuel market. For algae biodiesel, the competition is even more severe because dewatering from the algae-derived fuels costs greater than $3/gallon, which substantially exceeds the pre-tax price of petroleum alternatives (IEA, 2016 [24]; Uduman et al., 2010 [25]). The recent changes in the biodiesel policies in the EU and the US create additional challenges to biofuel production, and more investigation and development will be demanded more use of biodiesel (Bender, 1999) [26]. Direct or indirect biodiesel subsidies, which have been eliminated in many countries since 2010, may be acceptable to the policy makers only if the significant reduction of production costs of biodiesel is realized and exceptional contribution to the environment and health impacts is demonstrated (Petrou and Pappis, 2009) [27].

Above the current challenges of biodiesel fuel production, the changes of biodiesel policies and the reduction in the investment rate in the biofuel industry are making the future of biodiesel look even more unclear. The global investment in biofuel projects has dropped by 35% from 2014, and compared to 2008, it dropped more than 80% (UN, 2009 [28], 2016 [29]). There have been many abandoned projects by international oil companies (Hochman, 2014) [15].

2.4 Current Biodiesel Production

This section provides information on biodiesel production by continent. Biodiesel is a diesel equivalent derived from vegetable oils and animal fat. Large-scale production of biodiesel, however, is limited by the availability of low cost, sustainable feedstocks. In 2008, biodiesel production throughout the entire world was 14.7 billion liters.

Biodiesel is a supply option that has the potential to smooth the transition from a fossil fuel based energy paradigm to a new sustainable energy system. Although feedstock constraints limit the large scale development of biodiesel, it holds significant promise for small scale production. Besides, algae biodiesel production is expected to grow in the near future, and as a result, much of the feedstock constraints will be lifted. Information on the type of blends being proposed in different countries is included in Table 2.7.

Biodiesel is a superior standalone fuel that can be used in conventional engines with little modification. According to the Biofuel website

TABLE 2.7
Status of biodiesel production in different continents.

Continent	Biodiesel production
Asia	• According to the literature, Korea has planned for implementation of B3 since 2012. • Malaysia implemented B5 blending in 2008. • Philippines had plans for B1 in 2008 and B2 in 2011. • Thailand has planned for B10 started from 2012. Thailand has very aggressive plans for biodiesel. They implemented 3% blending in 2011 and are projecting 8.5 million liters of biodiesel production in 2012. Thailand started producing biodiesel in 2008 using palm oil. They produced 0.4 billion liters of biodiesel in 2008. • China is targeting 12 metric tons of biodiesel by 2012. China had started producing 0.1 billion liters of biodiesel using soybean and rapeseed in 2008. China is targeting 2 billion tons of biodiesel to be produced by 2020. • India has started producing biodiesel made from soybean and rapeseed oil. In 2008, India produced 0.02 billion liters of biodiesel.
South America	• Argentina started producing biodiesel using soybean in 2008. Argentina's production is 1.2 billion liters of biodiesel. Argentina has implemented blending protocol of B5 in 2012. • Bolivia implemented B2.5 in 2007 and has plans for B20 in 2015. • Brazil implemented B3 in 2008 and has plans for B5 in 2013. Brazil is producing 1.2 billion liters of biodiesel made from soybean. The use of biodiesel increased consumer price index by 3.27% and for other countries by 4.35%. • Chile implemented B5 in 2008. • Dominican Republic: Dominican Republic has plans for B2 in 2013. • Paraguay already implemented B5 in 2009. • Peru implemented B5 in 2011. • Uruguay planned using B5 in 2012.
Europe	• Germany implemented B5.25 in 2009 and has plans to achieve B6.25 through 2014. Germany produced 3.2 billion liters of biodiesel made from rapeseed in 2008. The consumption of biodiesel was 3,316,370 ton in 2010. Germany produced 3,349,868 ton rapeseed oil for biodiesel in 2010. It consumed 2,938,521 tons of biodiesel in 2010. 2,968,203 ton rapeseed oil for biodiesel manufacturing was produced in 2010. Germany has plans to increase

(*Continued*)

TABLE 2.7 (Cont.)

Continent	Biodiesel production
	the biodiesel consumption to 2,938,521 ton in 2015. Germany is also planning to increase their production of rapeseed oil to 2,968,203 ton in 2015. In 2005, Germany consumed 1,994,404 ton biodiesel in road transport. • Britain implemented B5 in 2010. The UK achieved 0.22 billion liters of biodiesel in 2008 using the rapeseed as the feedstock. The consumption in 2010 was 92,457 ton. Britain produced 93,391 ton rapeseed oil for biodiesel in 2010. The UK consumed 348,720 ton biodiesel in 2010. The UK produced 352,242 ton rapeseed oil for biodiesel in 2010. The UK has plans to increase biodiesel consumption to 348,720 ton in 2015. The UK is trying to increase rapeseed oil production to 352,242 ton in 2015. The UK consumed 29,040 ton biodiesel in road transport in 2005, which is equal to 37.51% of the total biofuel consumption in 2005. • Cyprus consumed 22 ton biodiesel in road transport in 2005, which is equal to the total biofuel consumption in 2005. Cyprus is targeting 36,158 ton consumption of biodiesel in 2015. Cyprus is planning to increase its production of rapeseed oil to 36,523 ton in 2015. • France is producing 2.06 billion liters of biodiesel made from rapeseed in 2008. France produced 1,185,182 ton rapeseed oil for biodiesel in 2010. France consumed 1,604,079 ton biodiesel in 2010. France produced 1,620,282 ton rapeseed oil for biodiesel in 2010. France has plans to increase its biodiesel consumption to 1,604,079 ton biodiesel in 2015. France is targeting to increase rapeseed oil production to 1,620,282 ton in 2015. France consumed 368,487 ton biodiesel in road transport in 2005, which is equal to 81.35% of the total biodiesel consumption in 2005. France has the production capacity of 243,000 ton of biodiesel in 2006. • Greece has the production capacity of 549,000 ton of biodiesel in 2006. Greece is targeting 290,996 ton consumption of biodiesel in 2015, and is planning to increase its production of rapeseed oil to 293,935 ton in 2015. • Italy achieved the production of 0.68 billion liters of biodiesel made from oilseeds in 2008. The consumption in 2010 was 606,231 ton. From rapeseed, Italy produced 612,354 ton rapeseed oil for biodiesel. Italy is aiming for an increase of rapeseed oil production to 1,724,722 ton in 2015. Italy consumed 200,000 ton biodiesel in road transport in 2005, which is equal to 97.37% of the total biodiesel consumption in 2005. Italy had the production capacity of 70,000 ton of biodiesel in 2006.

(*Continued*)

TABLE 2.7 (Cont.)

Continent	Biodiesel production
	• Portugal: Portugal has the production capacity of 140,000 ton of biodiesel in 2006. The country is targeting 220,116 ton consumption of biodiesel in 2015. Portugal is planning to increase its production of rapeseed oil to 220,340 ton in 2015.
	• Luxembourg consumed 638 ton biodiesel in road transport in 2005, which is equal to the total biofuel consumption in 2005. Luxembourg is targeting 107,260 ton consumption of biodiesel in 2015. Luxembourg is planning to increase its production of rapeseed oil to 108,343 ton in 2015.
	• The Netherlands consumed 2,683 ton biodiesel in road transport in 2005, which is equal to 81.35% of the total bio-fuel consumption in 2005. The Netherlands is targeting 12,108 ton consumption of biodiesel in 2015. The Netherlands is planning to increase its production of rapeseed oil to 12,230 ton in 2015.
	• Spain produced 0.24 billion liters of biodiesel using oilseeds in 2008. Spain consumed 1,164,314 ton biodiesel in 2010. Spain is trying to increase its rapeseed oil production to 1,176,075 ton in 2015. Spain consumed 26,970 ton biodiesel in road transport in 2005, which is equal to 17.5% of total biofuel consumption. Spain had the production capacity of 550,000 ton of biodiesel in 2006.
	• Poland: Poland achieved up to 0.31 billion liters of biodiesel production using rapeseed in 2008. Poland consumed 274,679 ton biodiesel in 2010. 277,454 tons of rapeseed oil was produced for biodiesel in 2010. Poland is planning to increase rapeseed oil production to 277,454 ton in 2015. Poland consumed 17,100 ton biodiesel in road transport in 2005, which is equal to 35.67% of the total biofuel consumption in 2005. Poland used to have the production capacity of 464,750 ton of biodiesel in 2006.
	• Finland has the production capacity of 170,000 ton of biodiesel in 2006. Finland is targeting 159,742 ton consumption of biodiesel in 2015. Finland is planning to increase its production of rapeseed oil to 161,355 ton in 2015.
	• Slovenia consumed 5,616 ton biodiesel in road transport in 2005, which is equal to the total biofuel consumption in 2005. Slovenia is targeting 69,185 ton consumption of biodiesel in 2015. Slovenia is planning to increase its production of rapeseed oil to 69,884 ton in 2015.
	• Czech Republic achieved 0.12 billion liters of biodiesel production in 2008 made from rapeseed oil. The country consumed 3,169 ton biodiesel in road transport in 2005, which is equal to the total biofuel consumption in 2005. The Czech Republic is targeting 3,782 ton consumption of biodiesel in 2015. The Czech Republic is

(*Continued*)

TABLE 2.7 (Cont.)

Continent	Biodiesel production
	planning to increase its production of rapeseed oil to 3,821ton in 2015. • Estonia is targeting 19,348 ton consumption of biodiesel in 2015. Estonia is planning to increase its production of rapeseed oil to 19,543 ton in 2015. • Lithuania consumed 7,500 ton biodiesel in road transport in 2005, which is equal to 92.04% of the total biofuel consumption in 2005. Lithuania is targeting 19,929 ton consumption of biodiesel in 2015. Lithuania is planning to increase its production of rapeseed oil to 20,131 ton in 2015. • Latvia consumed 2,890 ton biodiesel in road transport in 2005, which is equal to 83.92% of the total biofuel consumption in 2005. Latvia has the production capacity of 265,500 ton of biodiesel in 2006. Latvia is targeting 42,995 ton consumption of biodiesel in 2015. The country is planning to increase its production of rapeseed oil to 43,429 ton in 2015. • Malta consumed 788,000 ton biodiesel in road transport in 2005, which is equal to the total biofuel consumption in 2005. Malta is targeting 10,403 ton consumption of biodiesel in 2015. Malta is planning to increase its production of rapeseed oil to 10,509 ton in 2015. • Sweden achieved 0.11 billion liters of biodiesel made from rapeseed in 2008. The consumption in 2010 was 28,523 ton. Sweden produced 28,811 ton rapeseed oil for biodiesel in 2010. Sweden is planning to increase its rapeseed oil production to 20,664 ton in 2015. Sweden consumed 9,677 ton biodiesel in road transport in 2005, which is equal to 5.62% of the total biofuel consumption in 2005. • Ireland: Ireland consumed 1,193 ton biodiesel in road transport in 2005, which is equal to 99.07% of the total biofuel consumption in 2005. Ireland is targeting 208,702 ton consumption of biodiesel in 2015. Ireland is planning to increase its production of rapeseed oil to 210,810 ton in 2015. • Denmark achieved biodiesel production of 0.15 billion liters using oilseeds as the feedstock in 2008. Denmark is targeting 135,401 ton consumption of biodiesel in 2015. Denmark is planning to increase its production of rapeseed oil to 136,768 ton in 2015. The data from 2001 to 2006 of biodiesel suggest a considerable decrease in consumer price index by 40.9%. • Belgium has the production capacity for producing 795,264 ton of biodiesel in 2006. Belgium is targeting 324,946 ton consumption of

(Continued)

TABLE 2.7 (Cont.)

Continent	Biodiesel production
	biodiesel in 2015. Belgium is planning to increase its production of rapeseed oil to 328,228 ton in 2015.
	• Austria consumed 92,000 ton of biodiesel in road transport in 2005. biodiesel was the only fuel used in 2005. Austria has the production capacity for producing 200,000 ton of biodiesel in 2006. Austria is targeting 255,193 ton consumption of biodiesel in 2015. Austria is planning to increase its production of rapeseed oil to 257,770 ton in 2015.
North America	• Canada started biodiesel production in 2008 using oilseeds mostly from canola oil. On June 26, 2008, Parliament passed Bill C-33, which will require the use of 5% renewable content in gasoline by 2010 and 2% renewable content in diesel fuel by no later than 2012. Canada produced 0.1 billion liters of biodiesel in 2008.
	• The US has implemented B5 in New Mexico and B2 in Louisiana and Washington State. US produced 2.69 billion liters biodiesel made from soybean in 2008. biodiesel producers receive a tax credit of $1 per gallon of biodiesel produced from virgin oil, which could be either animal fats or oilseeds. Small producers with less than 16 million gallons of biodiesel per annum can receive a tax credit of $0.1 per gallon for the first 15 million gallons of production with a maximum tax credit being $1.5 million per year. A tax credit equal to 30% of the cost of alternatively brief fueling property can be claimed. The maximum amount of tax credit is $30,000 for business and $1,000 for individuals using alternative fuels such as B20. Projects generating energy from biodiesel can get grants of up to $500,000 and a long guarantee of up to 10 million dollars. In Montana, a tax credit equal to 15% of the cost to compensate for the depreciation of equipment in the construction and facilities to be used for the production of biodiesel. In Montana, producers receive 0.1 dollars for each gallon of biodiesel produced that represents an increase over the previous year's production. A refund of $0.02 per gallon is paid to the distributor a tax refund to the distributor and 0.01 dollar to the retailer for the previous quarter if biodiesel was produced entirely from the ingredients produced in the state of Montana. In Idaho, animal fats are a tax deduction for the distributor of biodiesel produced from oilseeds. It is provided in the form of a reduced tax rate. $0.225 per gallon is for B10 blend. The state of North Dakota provides a biodiesel tax credit of 10% per year up to 5 years for costs incurred to develop or modify a facility or blend biodiesel. The state of Illinois has a clean school bus program which provides the rebate of up to 80% or

(Continued)

TABLE 2.7 (Cont.)

Continent	Biodiesel production
	maximum $4,000 toward the purchase of alternative-fuel vehicles. The state gives exemption on sales tax on biodiesel blend of 10%. The use of biodiesel (using the data from 2001 to 2006) increased the consumer price index by 12.75%.

(http://biofuel.org.uk), biodiesel production by ranking regions is as shown in Table 2.8:

TABLE 2.8

Ranking regions by biodiesel production.

Region	Fuel Production (liters)	Major Feedstocks
Europe	7 Billion	Canola/Soybean/Barley
North America	3 Billion	Soybean
South America	4 Billion	Castor bean/Sunflower
Africa (including the Middle East)	Limited	Limited
Australia/Asia	1 Billion	Soybean/Jatropha/UCO/Coconut/Palm
Total	**15 Billion**	

2.5 Recent Headlines on Biodiesel Production

It is important to look at a few internet postings to get an idea of the trend in biodiesel area. The following news items are the excerpts picked up from Bloomberg News.

Recent Headlines:

1. London's Iconic Red Buses to Run on Biofuel Made From Old Coffee By Anna Hirtenstein, November 19, 2017, 7:01 PM EST: London's iconic red double-decker buses will soon run on a biofuel partially made from old coffee grounds. The fuel will be supplied by a demonstration project set up by Bio-bean Ltd., a London-based company that joined with Royal Dutch Shell Plc on the initiative. It will produce 6,000 liters (1,583 gallons) a year of the fuel. "It's got a high oil content, 20% oil by weight in the waste coffee grounds, so it's a really great thing to make biodiesel out of," said Arthur Kay, founder of Bio-bean, in a phone interview.

2. As public pressure mounts against using food for fuel, companies are increasingly focusing on biofuels made from waste such as UCO and inedible plants. Some crops such as corn and sugarcane are made into ethanol to be burned in engines, with sizable markets in some parts of the US and South America. Bio-bean has partnered with thousands of coffee shops across the UK, such as Costa Coffee Ltd. and Caffe Nero, to collect used grounds. The UK produces 500,000 tons annually, according to Kay. Caffe Nero's parent company is Italian Coffee Holdings Ltd., based in London. It will then be converted into biofuel at the company's factory in Cambridgeshire and blended with ordinary diesel with the finished product at 20%. It will then be shipped to a central tank where London buses refuel.

   The company also makes a solid biomass pellet and briquette to be used in home heating and in stoves, producing 50,000 tons per year. "It's also a good feedstock for our other products for instance because it is packed full of energy, they have a higher calorific content than wood," Kay said. Bio-bean was founded in 2013 and has received funding from the UK government, Shell and private investors. It is planning to expand throughout the UK and eventually to continental Europe and the US "We're basically looking for places where they drink a huge amount of coffee," Kay said. "Our primary expansion plans are based around where there are instant coffee factories."

3. Top Palm Oil Growers Go on Defensive Against EU Curb Threat By Anuradha Raghu, November 19, 2017, 10:33 PM EST Updated on November 20, 2017, 2:47 AM EST: As concerns about palm oil's sustainability simmer in Europe, the world's second-biggest grower is ramping up its defense of the most-consumed edible oil. Malaysian Prime Minister Najib Razak and Indonesian President Joko Widodo will this week discuss concerns that a resolution passed by the EU in April calling for tougher environmental standards for palm oil may hurt the industry. The two countries are the world's top palm oil producers, accounting for 85% of supply.

   "If such a resolution affects our exports, it will be a major blow," Mah Siew Keong, Malaysia's Plantation Industries and Commodities Minister, said in an interview in Kuala Lumpur. "Any form of discrimination is not acceptable, and we will be compelled to act if it's enforced to protect our own interest." The European Parliament's nonbinding resolution urged the bloc's executive arm to step up efforts to prevent deforestation as a result of palm oil production. The expansion of plantations in the two countries has seen farmers accused of illegally using slash-and-burn methods to clear land, destroying rainforests and habitats for animals, and causing a severe haze that can blanket parts of Asia. Indonesia has said it is ready to retaliate against further attempts to curb palm oil exports.

The EU is Malaysia's biggest export destination, accounting for about 13% of shipments of palm oil and palm-based products last year, according to the Malaysian Palm Oil Board. About 90% of Malaysia's biodiesel exports also go to Europe, Mah said. Mah met with 18 European ambassadors on Monday, according to a statement from the ministry. EU officials are consulting with stakeholders from palm oil producing countries and the resolution is being worked through European commissioners and the European Council before possible legislative measures are proposed, the statement cited Maria Castillo Fernandez, ambassador and head of the EU's delegation to Malaysia, as saying.

Indonesia's president has asked the EU to end discrimination against palm oil as it harms economic interests. Palm oil is the country's number one export commodity to the EU, representing 49% of the region's imports, according to the EU. The industry helps alleviate poverty, narrow the development gap and develops an inclusive economy, according to Widodo.

Biofuels Fight Is Brewing Between US and Brazil Over Ethanol By Mario Parker and Fabiana Batista, November 10, 2017, 1:54 PM EST: The US biofuels industry, fresh off a win against Big Oil, is lining up for a fight with Brazil. American ethanol producers said Thursday in a letter to US Trade Representative Robert Lighthizer that they're seeking Brazil's suspension from a trade program allowing duty-free imports into the US The move follows Brazil's decision in August to slap a 20% tariff on ethanol shipments from the US that exceed a 600 million-liter (158 million-gallon) annual quota.

The US ethanol lobby was buoyed last month by President Donald Trump's instruction to EPA Administrator Scott Pruitt to support the RFS, a law mandating the use of fuels such as corn-based ethanol and soy-based biodiesel. Trump's personal intervention came despite the objections of oil refiners.

The spat with Brazil also comes as the White House pursues a protectionist agenda in dealing with international trade. On Thursday, the Commerce Department set import duties on biodiesel from Indonesia and Argentina after US producers said they were harmed by unfair state subsidies given to competitors in those countries.

The letter to Lighthizer was signed by three industry groups: The Renewable Fuels Association, Growth Energy and the US Grains Council. Ethanol in the US is made from corn, making the industry an important part of the farm lobby. Trump visited ethanol plants in Iowa during his presidential campaign and told the state's voters he would stand by the biofuel if he was elected.

"It is fair to say that President Donald Trump's administration has been actively and constructively engaged in every step of the way of this process," Bob Dinneen, president of the Renewable Fuels Association, said Friday in a telephone interview. Dinneen said the government has already signaled to Brazil there will be consequences for imposing the tariff. American ethanol exports to the country dropped 54% in August from January levels, data from the US Energy Information Administration show. No one at the US Trade Representative or the Brazilian government immediately responded to requests for comment.

The ethanol groups are threatening to petition for Brazil's suspension designated country status under the Generalized System of Preferences trade program, which allows duty-free imports to promote economic growth in other countries. Brazil is the third-largest beneficiary under the program, with $2.2 billion of eligible trade in 2016, the groups said in their letter. Brazil's tariff violates the "spirit" of the program, they said.

Brazil didn't violate World Trade Organization rules, Eduardo Leao, executive director of Unica, a sugar and ethanol industry group, said in an interview Friday. The tariff is being levied only on the amount of ethanol that exceeds 600 million liters, which is the average amount of imports in the three years through 2016. "We made a courtesy to the Americans in taxing only what exceeds the quota." Leao said.

Exxon Quietly Researching Hundreds of Green Projects By Anna Hirtenstein, November 3, 2017, 1:00 AM EDT Corrected November 3, 2017, 1:04 PM EDT: One of the world's biggest oil companies is working on hundreds of low-carbon energy projects, from algae engineered to bloom into biofuels and cells that turn emissions into electricity.

The work by Exxon Mobil Corp. includes research on environmentally-friendly technologies in five to 10 key areas, according to Vice President of R&D Vijay Swarup. While any commercial breakthrough is at least a decade away, Exxon's support for clean energy suggests the world's most valuable publicly-traded oil company is looking toward the possibility of a future where fossil fuels are less dominant.

While Exxon has discussed some of its research before and runs advertisements about its work in algae, the remarks from Swarup are the first indication of the breadth of the oil company's interests in alternative energies. It's part of the $1 billion a year Exxon spends on research worldwide and the $8 billion it has spent since 2000 researching, developing and deploying low-carbon technologies. "These areas are massively challenging, and if we can solve those, they will have huge impacts on our business," said Swarup in a phone interview. "We bring more than money. We bring the science, the commitment to research."

Exxon didn't disclose the exact amount its spending on the green technologies. The broader investments it has made since the beginning of the century also include things like managing methane emissions from oil wells, on co-generation and on making its plants more efficient.

The company joins a growing list of oil majors hedging against the wider adoption of renewables, which could displace some 8 million barrels of crude demand a day, according to Bloomberg New Energy Finance. Some companies, like France's Total SA, have made acquisitions to enter the business. Others, like Royal Dutch Shell Plc, are using experiences from running offshore rigs to develop wind farms in the North Sea. Based in Irving, Texas, Exxon said its approach is different because it is focusing on science, Swarup said. It has joined with about 80 universities and is collaborating with smaller companies on research. Projects it's working on with respect to biofuels include algae biofuels. Exxon is planning to harvest algae in ponds or oceans around the world and process it into biofuel for regional distribution. Swarup expects that it will first be blended with diesel and jet fuel, but the goal is to eventually sell a 100% algae-derived fuel.

Biodiesel made from agricultural waste. The company is working with Renewable Energy Group Inc. to use microbes to convert inedible crop residue like corn husks into biofuels. The two companies began their collaboration in 2016 and recently extended their joint research program.

"We are still 10 plus years away" for the algae biofuels to be deployed at scale, according to Swarup, who said the company's been focusing research on algae for eight years.

4. California Cars Are Running on Restaurant Grease By Robert Tuttle, October 23, 2017, 5:36 PM EDT Updated on October 24, 2017, 7:00 AM EDT: California's battle against climate change is being fought more fiercely in fast food restaurants than in Tesla Inc.'s car factory in Fremont. Seven years after the Golden State began offering credits to producers of low-carbon fuels, cities and companies across California are using diesel brewed from fats and oils to fuel everything from fire trucks to United Parcel Service Inc. delivery vehicles. Now, the value of the credits exceed those from electric vehicles fourfold and are second only to ethanol.

   The company that's benefited most from California's embrace of renewable diesel is based 6,000 miles away in Helsinki. Neste Oyj started sending tankers of the fuel from its refineries in Singapore and Europe around 2012. It's now the biggest supplier, according to Ezra Finkin, policy director at the Diesel Technology Forum, a Frederick, Maryland-based advocacy group.

The market "is definitely growing," Dayne Delahoussaye, Neste's head of North American public affairs, said in a phone interview from Houston. "Renewable diesel has become very popular with the refining community as a good tool to meet obligations."

Renewable diesel generated almost 628,000 metric tons of credits in the fourth quarter of last year, up from about 6,000 in 2011, state data show. The credits, which sold at a six-month high of $91.74 per metric ton in early October, are poised to surge as the state accelerates its carbon cuts to meet its goal of reducing emissions to 30-year-old levels by decade's end. The price will more than double to $215 by 2019, Irvine, California-based Stillwater Associates LLC said in a June report.

Refiners and other purchasers of the credits have paid almost $650 million for them over the past year, costs that are passed onto consumers at the pump. The credits will add 15 to 20 cents a gallon to the cost of fuel over the next two years, Leigh Noda, senior associate at Stillwater Associates, said in a phone interview. "Ultimately, these programs are designed to subsidize the price of the biofuel suppliers."

In recent years, cities such as San Francisco, Oakland, and San Diego, as well as Sacramento County, have transitioned to using renewable diesel to power buses, fire engines and other city vehicles. Alphabet Inc.'s shuttle buses in Silicon Valley also burn it, and UPS said two years ago that it would buy 46 million gallons of the fuel to run its fleet of delivery trucks.

2.6 Renewable Fuel Standard (RFS)

The RFS program was created under the Energy Policy Act of 2005, which amended the Clean Air Act (CAA). The EISA of 2007 further amended the CAA by expanding the RFS program. EPA implements the program in consultation with the US Department of Agriculture and the Department of Energy.

The RFS program is a national policy that requires a certain volume of renewable fuel to replace or reduce the quantity of petroleum-based transportation fuel, heating oil or jet fuel. The four renewable fuel categories under the RFS are:

- Biomass-based diesel
- Cellulosic biofuel
- Advanced biofuel
- Total renewable fuel

The 2007 enactment of EISA significantly increased the size of the program and included key changes, including:

- Boosting the long-term goals to 36 billion gallons of renewable fuel
- Extending yearly volume requirements out to 2022
- Adding explicit definitions for renewable fuels to qualify (e.g., renewable biomass, GHG emissions)
- Creating grandfathering allowances for volumes from certain existing facilities
- Including specific types of waiver authorities

Volumes provided in the statute are listed in Table 2.9 (EPA):

TABLE 2.9

Volume standards as set forth in EISA.

Year	Cellulosic Biofuel	Biomass-Based Diesel	Advanced Biofuel	Total Renewable Fuel	Conventional Biofuel
2009	NA	0.50	0.60	11.10	10.5
2010	0.10	0.65	0.95	12.95	12.0
2011	0.25	0.80	1.35	13.95	12.6
2012	0.50	1.00	2.00	15.20	13.2
2013	1.00	*	2.75	16.55	13.8
2014	1.75	*	3.75	18.15	14.4
2015	3.00	*	5.50	20.50	15.0
2016	4.25	*	7.25	22.25	15.0
2017	5.50	*	9.00	24.00	15.0
2018	7.00	*	11.00	26.00	15.0
2019	8.50	*	13.00	28.00	15.0
2020	10.50	*	15.00	30.00	15.0
2021	13.50	*	18.00	33.00	15.0
2022	16.00	*	21.00	36.00	15.0

* Statute sets 1 billion gallons minimum, but EPA may raise the requirement.
Note: There is no statutory volume requirement for conventional biofuel. The conventional volumes in the table are calculated (total – advanced) and are certain biofuels that do not qualify as advanced.

References

1. US General Accounting Office, (2000), Financial Report of the United States Government in 2000. https://www.gao.gov/special.pubs/00frusg.pdf
2. Solomon, B.D., Barnes, J.R., and Halvorsen, K.E., (2007). Grain and cellulosic ethanol: history, economies, and energy policy. *Biomass and Bioenergy*, *31*(6), pp. 416–425.
3. Rubin, O.D., Carriquiry, M., Hayes, D.J., (2008). *Implied Objectives of US Biofuel Subsidies. Center for Agricultural and Rural Development*, Working Paper 08-WP 459:1–31.
4. Gecan, R., Johansson, R., (2010). *Using Biofuel Tax Credits to Achieve Energy and Environmental Policy Goals.* US Congressional Budget Office, July.
5. Joseph, K., (2017). *Argentina biofuels annual. USDA Foreign Agricultural Service, GAIN Report Number*, approved by Lazaro Sandoval.
6. Colares J., (2008). A brief history of Brazilian biofuel legislation. *Syracuse Journal of Law Commerce*, *35*(2), pp. 101–116. Contract 59-2171-1-6-057-0, USDA, ARS, Peoria, IL.
7. de Almeida, E.F., Bomtempo, J.V. and de Souzae Silva, C.M., (2008). *Theperformanceof Brazilian Biofuels: Aneconomic, Environmental and Social Analysis.*

Published in Biofuels - Linking Support to Performance by the OECD/ITF, pp. 151–188.

8. Bognar, J., Mondou, M. and Skogstad, G., (2017). *Best Practices for Biofuel Policy* (accessed on March 16, 2018). http://biofuelnet.ca/wp-content/uploads/2017/09/2017
9. Barros S., (2009). *Brazil Biofuels Annual—Biodiesel Annual Report. USDA Foreign Agricultural Service, GAIN Report Number BR9009*, July 31.
10. Barros, S., (2017). *Brazil biofuels annual. USDA Foreign Agricultural Service. Global Agricultural Information Network, GAIN Report Number BR17006*, approved by Chanda Berk.
11. UFOP, (2016/2017). *Report on Progress and Future Prospects – Excerpt from the UFOP Annual Report, Biodiesel.* www.ufop.de/files/5115/1309/0426/UFOP-Biodiesel_2016-2017_EN.pdf
12. Pires, A. and Schechtman, R., (2010). International Biofuels Policies. In E.L. Sousa & I.C. Macedo, eds. *Ethanol and Bioelectricity Sugarcane in the Future of the Energy Matrix*, São Paulo: UNICA, pp. 191–224.
13. Dong, F., (2007). *Food Security and Biofuels Development: the Case of China.* Briefing Paper 07- BP-52, Centre for Agricultural and Rural Development, Iowa State University, October.
14. MPPU, Archieve/January (2018). *Fao Trade and Markets Division–Oilseeds desk* www.fao.org/fileadmin/templates/est/COMM_MARKETS_MONITORING/Oilcrops/Documents/archive_Jan_2018.xlsm.
15. Hochman, G., (2014). Biofuels at a crossroads. *Choices, 29*, pp. 1–5.
16. Canakci, M. and Sanli, H., (2008). Biodiesel production from various feedstocks and their effects on the fuel properties. *Journal of industrial microbiology & biotechnology, 35*(5), pp. 431–441.
17. Demırbas, A., (2017). The social, economic, and environmental importance of biofuels in the future. *Energy Sources, Part B: Economics, Planning, and Policy, 12* (1), pp. 47–55.
18. Rocha, M.V.P., de Matos, L.J.B.L., de Lima, L.P., Da Silva Figueiredo, P.M., Lucena, I.L., Fernandes, F.A.N. and Gonçalves, L.R.B., (2014). Ultrasound-assisted production of biodiesel and ethanol from spent coffee grounds. *Bioresource Technology, 167*, pp. 343–348.
19. Ambat, I., Srivastava, V. and Sillanpää, M., (2018). Recent advancement in biodiesel production methodologies using various feedstock: a review. *Renewable and Sustainable Energy Reviews, 90*, pp. 356–369.
20. Chhetri, A.B., Tango, M.S., Budge, S.M., Watts, K.C. and Islam, M.R., (2008). Non-edible plant oils as new sources for biodiesel production. *International Journal of Molecular Sciences, 9*(2), pp. 169–180.
21. Omidvarborna, H., Kumar, A. and Kim, D.S., (2015). NOx emissions from low-temperature combustion of biodiesel made of various feedstocks and blends. *Fuel Processing Technology, 140*, pp. 113–118.
22. Mata, T.M., Martins, A.A. and Caetano, N.S., (2010). Microalgae for biodiesel production and other applications: a review. *Renewable and Sustainable Energy Reviews, 14*(1), pp. 217–232.
23. IEA (International Energy Agency), (2007). *Biodiesel Statistics. IEA Energy Technology Essentials.* Paris: OECD/IEA.
24. IEA (International Energy Agency) World Energy Outlook, (2016). IEA/OECD: Paris, France.

25. Uduman, N., Qi, Y., Danquah, M.K., Forde, G.M. and Hoadley, A., (2010). Dewatering of microalgal cultures: a major bottleneck to algae-based fuels. *Journal of renewable and sustainable energy*, 2(1), p. 012701.
26. Bender, M., (1999). Economic feasibility review for community-scale farmer cooperatives for biodiesel. *Bioresource Technology*, *70*, pp. 81–87.
27. Petrou, E.C. and Pappis, C.P., (2009). Biofuels: a survey on pros and cons. *Energy & Fuels*, *23*(2), pp. 1055–1066.
28. UN Environment Programme and Bloomberg New Energy Finance (UNEP-BNEF), (2009). Global trends in sustainable energy investments in 2009. Frankfurt School-UNEP Centre/BNEF: Frankfurt, Germany.
29. UN Environment Programme and Bloomberg New Energy Finance (UNEP-BNEF), (2016). Global trends in sustainable energy investments in 2016. Frankfurt School-UNEP Centre/BNEF: Frankfurt, Germany.
30. Major biofuel producers by region, Biofuels, the fuel of future. http://biofuel.org.uk/major-producers-by-region.html.

3

Biodiesel Policy in India

Aashti Hamid
Consultant, Vadodara, Gujarat

3.1 Introduction

Economic development in India has raised millions of people out of poverty and brought about the modernization of society. Economic ambition though has not been reached without costs. India has become more reliant of imports of energy that affects energy security. Pollution from industry, transport and traditional cook stoves affects air quality and is increasing emissions of greenhouse gases and contributing to climate change. The highest demand for energy comes from industry, followed by the transportation sector that consumed about 16.9% (36.5 of oil equivalent) of the total energy (217 million tons) in 2005–2006. It is estimated that the demands of diesel and gasoline will rise from 80.4 MMT and 26.1 MMT, respectively, in FY 2017–2018 to 110 MMT and 31.1 MMT by the year 2021–2022, if the current situation prevails. This growth will only escalate over the next several years since India's vehicular population is expected to grow by 10–12% per annum. Hence, securing a long-term supply of energy sources and prioritizing development are critical to ensure that the country's future energy requirements will be met. Therefore, India initiated biofuel production nearly a decade ago to reduce its dependence on imported oil and thus improve energy security and is now one of the largest producers of Jatropha oil. The country began 5% ethanol blending (E5) pilot program in 2001 and formulated the National Mission on Biodiesel in 2003 to achieve 20% biodiesel blends by 2011–12. Similar to many countries around the world, India's biofuel programs experienced setbacks, primarily because of supply shortages and global concerns over food security. India's National Policy on Biofuels in 2009 proposed a nonmandatory target of a 20% blend for both biodiesel and ethanol by 2017, and outlines a broad strategy for the biofuels program and policy measures to be considered to support the program. Several ministries in India are currently involved in the promotion, development, and policy making for the biofuel sector.

- The Ministry of New and Renewable Energy is the overall policymaker, promoting the production, research and development of biofuels.

- The Ministry of Petroleum and Natural Gas is responsible for marketing biofuels in the development and implementation of pricing and procurement policy.
- The Ministry of Agriculture's role is that of promoting research and development for the production of biofuel feedstock crops.
- The Ministry of Rural Development is specially tasked to promote Jatropha plantations on wastelands.
- The Ministry of Science & Technology supports research in biofuel crops, specifically in the area of biotechnology.

In view of the multiple departments and agencies involved, a National Biofuel Coordination Committee (NBCC) headed by the Prime Minister was set up to provide high-level coordination and policy guidance/review on different aspects of biofuel development, promotion and utilization.

3.2 India's Biofuel Policy

In 1948, the Power Alcohol Act heralded India's recognition of blending petroleum fuel with ethanol. The main objective was to use ethanol from molasses to blend with petroleum fuel to bring down the price of sugar, trim wastage of molasses and reduce dependence on petroleum imports. Subsequently, the Act was repealed in 2000, and in January 2003, the Government of India launched the Ethanol Blended Petrol (EBP) Program in nine States and four Union Territories promoting the use of ethanol for blending with gasoline and the use of biodiesel derived from nonedible oils for blending with petroleum diesel (5% blending). In April 2003, the National Mission on Biodiesel launched by the Government identified *Jatrophacurcas* as the most suitable tree-borne oilseed for biodiesel production.

Due to the ethanol shortage during 2004–05, the blending mandate was made optional in October 2004, but was resumed in October 2006 in 20 States and 7 Union Territories in the second phase of EBP Program. These ad-hoc policy changes continued until December 2009, when the Government came out with a comprehensive National Policy on Biofuels formulated by the Ministry of New and Renewable Energy (MNRE), calling for blending at least 20% biofuels with diesel and petrol by 2017.

3.2.1 National Policy on Biofuels-2018: An Overview

The Union Cabinet, chaired by the Prime Minister Shri Narendra Modi has approved National Policy on Biofuels – 2018.

Salient Features:

i. The Policy categorizes biofuels as "Basic Biofuels" viz. First Generation (1G) bioethanol & biodiesel and "Advanced Biofuels" – Second Generation (2G) ethanol, Municipal Solid Waste (MSW) to drop-in fuels, Third Generation (3G) biofuels, bio-CNG (compressed natural gas), and so on to enable extension of appropriate financial and fiscal incentives under each category.

ii. The Policy expands the scope of raw material for ethanol production by allowing use of sugarcane juice, sugar-containing materials like sugar beet, sweet sorghum, starch-containing materials like corn, cassava, damaged food grains like wheat, broken rice, and rotten potatoes, unfit for human consumption for ethanol production.

iii. Farmers are at a risk of not getting an appropriate price for their produce during the surplus production phase. Considering this, the policy allows the use of surplus food grains for production of ethanol for blending with petroleum fuel with the approval of National Biofuel Coordination Committee.

iv. With a thrust on Advanced Biofuels, the policy indicates a viability gap funding scheme for 2G ethanol bio refineries of Rs.5,000 crore, approximately 700 million US dollars, in 6 years in addition to additional tax incentives, higher purchase price as compared to 1G biofuels.

v. The Policy encourages setting up of supply chain mechanisms for biodiesel production from nonedible oilseeds, Used Cooking Oil, short gestation crops.

vi. Roles and responsibilities of all the concerned ministries/departments with respect to biofuels have been captured in the policy document to synergize efforts.

3.2.2 Expected Benefits

3.2.2.1 Reduce Import Dependency

One crore (10^7) litter of E10 saves Rs. 28 crore (4 million US dollars) of forex at current rates. The ethanol supply in year 2017–2018 is likely to see a supply of around 150 crore liters (1.5 billion liters) of ethanol, which will result in savings of over Rs. 4000 crore ($600 million) of forex.

3.2.2.2 Cleaner Environment

One crore lit of E10 saves around 20,000 ton of CO_2 emissions. For the ethanol supply year 2017–2018, there will be less emissions of CO_2 to the tune of 3 million tons. By reducing crop burning & conversion of agricultural residues/wastes to biofuels there will be further reduction in greenhouse gas emissions.

3.2.2.3 Health benefits

Prolonged reuse of cooking oil for preparing food, particularly in deep-frying is a potential health hazard and can lead to many diseases. Therefore, utilizing used cooking oil as a potential feedstock for biodiesel and its use for making biodiesel will prevent diversion of used cooking oil in the food industry.

3.2.2.4 MSW Management

It is estimated that, annually 62 MMT of municipal solid waste is generated in India. There are technologies available which can convert waste/plastic, MSW, to drop in fuels. One ton of such waste has the potential to provide around 20% of drop in fuels.

3.2.2.5 Infrastructural Investment in Rural Areas

It is estimated that, one 100,000 L/day biorefinery will require around Rs.800 crore capital investment ($113 million). At present Oil Marketing Companies are in the process of setting up 12 2G bio refineries with an investment of around Rs. 10,000 crore ($1.4 billion). Further addition of 2G bio refineries across the country will spur infrastructural investment in the rural areas.

3.2.2.6 Employment Generation

One 100klpd 2G bio refinery can contribute 1200 jobs in Plant Operations, Village Level Entrepreneurs, and Supply Chain Management.

3.2.2.7 Additional Income to Farmers

By adopting 2G technologies, agricultural residues/waste which otherwise are burnt by the farmers can be converted to ethanol and can fetch a price for these waste if a market is developed for the same. In addition, farmers are at a risk of not getting appropriate prices for their produce during the surplus production phase. Thus, conversion of surplus grains and agricultural biomass can help in price stabilization.

3.3 India's Biodiesel Policy – 2018

India's National Bio-Diesel policy is driven by the fact that energy is a critical input for socio-economic development of our country. The energy strategy aims to meet the government's recent ambitious announcements in the energy domain such as electrification of all census villages by 2019, 24/7 electricity & 175 GW of renewable energy capacity by 2022, reduction in energy emissions intensity by 33%–35% by 2030 and to increase the share of non-fossil fuel based capacity in the electricity mix above 40% by 2030. Even if there is likely expansion in the energy contribution of oil, gas, coal, renewable resources, nuclear and hydro in the coming decade, fossil fuels will continue to occupy a significant share in the energy basket. However, India is endowed with abundant renewable energy resources therefore; their use needs to be encouraged in every possible way.

In addition, the crude oil price has been fluctuating in the world market. Such fluctuations are straining various economies the world over, particularly those of the developing countries. Road transport sector accounts for 6.7% of India's Gross Domestic Product (GDP). Currently, diesel alone meets an estimated 72% of transportation fuel demand followed by petrol at 23% and balance by other fuels such as CNG, LPG, and so on for which the demand has been steadily rising. Provisional estimates have indicated that crude oil required for indigenous consumption of petroleum products in FY 2017-18 is about 210 MMT. The domestic crude oil production is able to meet only about 17.9% of the demand, while the rest is met from imported crude. India's energy security will remain vulnerable until alternative fuels to substitute/supplement petro-based fuels are developed based on indigenously produced renewable feedstock. To address these concerns, Government has set a target to reduce the import dependency by 10% by 2022. Therefore, Government of India has prepared a road map to reduce the import dependency in the oil & gas sector by adopting a five-pronged strategy, which includes (1) increasing domestic production, (2) adopting biofuels and renewables, (3) increasing energy efficiency norms, (4) improvement in the refinery processes, and (5) demand substitution. This envisages a strategic role for biofuels in the Indian Energy basket.

Biofuels are derived from renewable biomass resources and wastes such as plastic, municipal solid waste (MSW), waste gas and so on, and therefore, they seek to provide a strategic advantage to promote sustainable development. This supplements conventional energy sources in meeting the rapidly increasing requirements for transportation fuels associated with high economic growth, as well as in meeting the energy needs of India's vast rural population. India believes that biofuel can increasingly satisfy India's growing energy needs in an environmentally benign and cost effective manner; reducing dependence on the import of fossil fuels, and thereby, providing a higher degree of national energy security. The energy policy says that Indian approach to biofuel is different from the current international approaches that could lead to conflict with food security. It is based solely on nonfood feedstocks to be raised on degraded or wastelands that are not suited to agriculture, thus avoiding a possible conflict of fuel versus food security. Over the last decade, Government of India has undertaken multiple interventions in promoting and revamping biofuels in the country through various structured programs like Ethanol Blended Petrol Program, National Biodiesel Mission, Biodiesel Blending Program, and so on, which have positively impacted the overall biofuels program in the country.

3.3.1 Strategy and Approach

Government is adopting a multi-pronged approach to promote and encourage use of biofuels by:

1. Blending ethanol in petrol through the Ethanol Blended Petrol (EBP) Program using ethanol derived from multiple feedstocks.
2. Development of Second Generation (2G) ethanol technologies and its commercialization.
3. Blending biodiesel in petroleum diesel through Biodiesel Blending Program exploring multiple feedstocks including straight vegetable oil in stationery, low RPM engines.
4. Focus on drop-in fuels produced from MSW, industrial wastes, biomass, and so on.
5. Focus on advanced biofuels including bio-CNG, biomethanol, DME, biohydrogen, bio-jet fuel, and so on.

In India, bioethanol can be produced from multiple sources like sugar containing materials, starch containing materials, celluloses and lignocelluloses material including petrochemical routes. However, the present policy of Ethanol Blended Petrol (EBP) Program allows bioethanol to be procured from non-food feed stock like molasses, celluloses and lignocelluloses material including petrochemical routes. Similarly, biodiesel can be produced from any edible/nonedible oilseeds, used cooking oil (UCO), animal tallow, acid oil, algal feedstock, and so on. However, biodiesel coming for the blending program is presently being manufactured from imported sources like palm stearin.

Cultivators, farmers and landless laborers and so on are encouraged to grow variety of different fuel crops as well as oil seeds on their marginal lands, as inter-crop and as second crop wherever only one crop is raised by them under rain-fed conditions so as to provide the feedstock for biodiesel and bioethanol. Corporates are also enabled to undertake plantations through contact farming by involving farmers, co-operatives and self-help groups, and so on. Village Panchayat (Indian local government) and communities play a very crucial role in augmenting indigenous feedstock supplies for biofuel production. In cases relating to usage of wastelands, degraded or fallow land in forest and nonforest areas for feedstock generation, local communities from Gram Panchayats (local government council)/talukas are encouraged for the plantation of nonedible oil seeds bearing trees/crops such as Pongamiapinnata (Karanja), Melia azadirachta (Neem), JatropaCarcus, CallophylumInnophylum, Simaroubaglauca, Hibiscus cannabbinusetc, short rotation crops such as sweet sorghum and energy grasses, for example, Miscanthusgigantum, switchgrass (Panicumvigratum), giant reed (Arundodonax), and so on. The support also comes through a Minimum Support Price (MSP) mechanism for oilseeds with a provision of periodic revision so as to ensure a fair price to the farmers. The National Biofuels Co-Ordination Committee is entrusted with this task. The Statutory Minimum Price (SMP) mechanism prevalent for sugarcane procurement was examined and has been used for oilseeds, utilized in production of biodiesel. Payment of SMP is the responsibility of the biodiesel processors.

3.3.2 Biodiesel Blending Program

The overall blending percentage of biodiesel in petroleum diesel has been less than 0.5% in the country due to constraints pertaining to feedstock availability. Moreover, the biodiesel controlled by the blending program is manufactured from imported sources. Thus, ensuring domestic raw material for biodiesel production is integral for long-term success of this program. It is also recommended to permit blending up to certain prescribed levels initially, to review and moderate the blending levels periodically as per the availability of biodiesel and to make it mandatory in due course. A proposal is made to form a National Registry that notes the feedstock availability, processing facilities and off-take and maintains the necessary data. This data is then be used for decision-making. In order to take care of fluctuations in the availability of biofuels, oil manufacturing companies (OMCs) are permitted to bank the surplus quantities left after blending of biodiesel and bioethanol in a particular year, and to carry it over forward to the subsequent year when there may be a shortfall in their availability to meet the prescribed levels. Blending follows a protocol and certification process in accordance to the BIS (Bureau of Indian Standards) specification and standards.

In-house produced used/waste cooking oil (UCO/WCO) offers potential to be a source of biodiesel production. However, the same is marred by diversion of UCO to edible stream through various small eateries, vendors and traders. Focus is laid upon laying down the stringent norms for preventing UCO from entering into the food stream and developing a suitable collection mechanism to augment its supply for biodiesel production.

3.3.3 Distribution and Marketing

The responsibility of storage, distribution, and marketing of biofuels is with OMCs. This is carried out through their existing storages and distribution infrastructure and marketing networks. The entire value chain comprising production of oil seeds, extraction of bio-oil, its processing, blending, distribution, and marketing is taken into account for determination of biodiesel purchase price. The Minimum Purchase Price (MPP) for biodiesel by OMCs is linked to the prevailing retail diesel prices. The MPPs, both for biodiesel and bioethanol, are determined by the Biofuel Steering Committee and decided by the National Biofuel Coordination Committee. In the event of diesel or petroleum price falling below the MPPs for biodiesel and bioethanol, OMCs are duly compensated by the Government.

3.3.4 Financing

Plantation of nonedible oil-bearing plants, the setting up of oil expelling/extraction and processing units for production of biodiesel and creation of any new infrastructure for storages and distribution is declared as a priority sector for lending from financial institutions and banks. National Bank of Agriculture and Rural Development (NABARD) and other Public Sector Banks are being encouraged by the Government to provide funding, financial assistance through soft loans to farmers for plantations. Biofuel technologies and projects are allowed 100% foreign equity through an automatic approval route to attract Foreign Direct Investment (FDI), provided biofuel is for domestic use only, and not for export. As biofuels are derived from renewable biomass resources, they are also eligible for various fiscal incentives and concessions available to the New and Renewable Energy Sector from the Central and State Governments. Bioethanol enjoys concessional excise duty of 16% and biodiesel is exempted from excise duty. No other central taxes and duties were proposed to be levied on biodiesel and bioethanol. Custom and excise duty concessions are provided on plant and machinery for production of biodiesel or bioethanol, as well as for engines run on biofuels for transport, stationary and other applications, if these are not manufactured indigenously.

3.3.5 Major Funding and Implementing Agencies

To meet the ambitious national target for a green path to growth, R&D and demonstration, India partnered with other 22 member countries in Paris, on the "Mission Innovation (MI)" launched in November 2015 to drive forward clean energy innovation on a global scale. Mission innovation aims to escalate the pace of global clean energy innovation by doubling their governments clean energy research and development (R&D) investments over five years (US$30 billion per year from the current level of US$15billion per year), while encouraging greater levels of private sector investment in transformative clean energy technologies. It is aimed that these additional resources will dramatically accelerate the availability of the advanced technologies. The focus areas of MI and funding levels are given in Tables 3.1 and 3.2.

TABLE 3.1

Showing focus areas of MI and their share of overall funding in percentages.

Focus areas of MI	Share of overall funding (%)
Industry & buildings	5%
Vehicles & other transportation	15%
Biobased fuels and energy	**5%**
Solar, wind, and other renewables	40%
Hydrogen and fuel cells	5%
Cleaner fossil energy	10%
Electricity grid	10%
Energy storage	5%
Basic energy research	5%

(These are indicative figures and the actual percentage spend in each area will be decided based upon the discussions and firming up of project plans in each area.)

TABLE 3.2

Current Investment Status and Targets as per Mission Innovation Objectives

S. No		(Rs. in Crores)	$ (Rounded off)
1	Baseline Amount	470	$72.00 million
2	Baseline Year(s)	2014–15	
3	Doubling Target	940	145.00 million US$
4	Doubling Year	2019–20	
5	First-Year Amount	475	73.00 million US$

Currently in India, for R&D and demonstration projects, grants are being provided to academic institutions, research organizations, specialized centers, and industry. Strengthening of existing R&D centers and setting up of specialized centers in high technology areas is also considered. International scientific and technical cooperation in the area of clean energy is established in accordance to the national priorities, socioeconomic development strategies and goals. Modalities of such cooperation include joint research and technology development, field studies, pilot-scale plants, and demonstration projects with active involvement of research institutions and industry on either side. Appropriate bilateral and multilateral cooperation programs for sharing of technologies and funding are developed. Participation in international partnerships and exchange of scientists, wherever necessary, is also explored.

Several key government departments and research agencies are engaged in the work of clean energy but the major contributors are:

- Ministry of Science & Technology (specifically Department of Biotechnology); Department of Science and Technology, Council of Scientific and Industrial Research.
- Universities and technical institutes like IIT's through Ministry of Human resource Development.
- Ministry of New and Renewable Energy.
- Private Sector research institutes.
- Research institutes funded by public sector companies.

The National Policy on Biofuels mandates a blending of biofuels of approximately 20% of biofuels in transport fuels. This is an ambitious target and, therefore, Department of Biotechnology (DBT), Ministry of Science & Technology is majorly working on development of second-generation biofuels. DBT has established a large network of more than 100 scientists in the country, who are working to realize the goals set in National Biofuel Policy. Efforts are being made to support R&D towards the development of cost-effective next generation biofuels like algal-biodiesel, cellulosic ethanol, biobutanol, and biohydrogen.

So far, four Bioenergy Centers are established by the Department, specifically to strengthen the research base in the country in biofuel area and to promote translation of process and technologies from research to scale up and commercialization. These centers have the state-of-the-art facility and research teams are working in interdisciplinary areas related to Bioenergy. Various technologies including cellulosic ethanol have been developed.

Following four DBT-Bioenergy Centers exist with specific goals and targets set by Department aligned to National Biofuel Policy:

i. DBT-ICT Centre for Energy Biosciences, Matunga, Mumbai.
ii. DBT-IOC Centre for Advanced Bioenergy, Faridabad.
iii. DBT-ICGEB Centre for Advanced Bioenergy, New Delhi.
iv. DBT-Pan-IIT Center for Bioenergy (participating 5 IITs).

Department of Science & Technology has also supported several programs to evaluate the effect of biofuel use in automobiles in reducing emissions. Large programs are also undertaken with Indian Railways for the use of biodiesel in traction engines. Jatropha Micro Mission and Algal Network programs have also been successfully completed by the researchers with a collection of good feedstock material for biodiesel production. These days government is also giving special focus to algal biomass as a feedstock for biodiesel.

3.4 Developments in National Biodiesel Mission

A summary is given below:

Date	Action
April, 2003	Demonstration phase 2003 to 2007: Ministry of Rural Development appointed as nodal ministry to cover 400,000 hectares under Jatropha cultivation. This phase also proposed nursery development, establishment of seed procurement and establishment centers, installation of trans-esterification plant, blending and marketing of biodiesel.
October, 2005	The MoPNG announced biodiesel purchase policy in which OMC would purchase biodiesel from 20 procurement centers across the country to blend with high speed diesel with effect from January 2006. Purchase price set at INR 26.5 per liter.
CY, 2008	Self-Sustaining Execution phase 2008 to 2012: Target was to produce sufficient biodiesel for 20% blending by end of 2011 in five year plan, 2008–12. But, lack of large scale plantation, conventional low yielding Jatropha cultivars, seed collection and extraction infrastructure, buy-back arrangement, capacity and confidence building measures among farmers impeded the progress of this phase.
CY, 2010	An estimated 0.5 million hectares was covered under Jatropha cultivation of which two third plant population was new plantation and took two to three years to mature.
CY, 2011	No additional wastelands were brought under Jatropha cultivation except for few captive plantations managed by OMCs.
CY, 2012	The production of biodiesel from Jatropha seeds remained commercially insignificant as no biodiesel from Jatropha was procured by oil marketing companies for blending in the past three to four years.
CY, 2013	Biodiesel production from multiple feedstocks (crude oil, used cooking oils, animal fats and so on) was an economically viable option left with the producers. Industry sources claimed that small to medium scale industries were the major buyers of biodiesel who blended it with conventional diesel. The average purchase price of biodiesel in India then was around INR 45–48 ($0.64–0.68) per liter (including freight) and seemed viable for blending since regular diesel was selling at a price premium of 18–20% over biodiesel.
CY, 2014	Spotlight came on the industry engagement with tree borne oilseeds (neem, pongamia, mahua, and kusum) as an alternative to Jatropha for biodiesel production as the seed yield from Jatropha plantation (on pilot scale) were observed to be significantly lower than stipulated. Therefore, the cost of production of biodiesel from Jatropha seed became too high and producers had little incentive to produce at peak capacity. In October 2014, the Government of India (GOI) deregulated the diesel prices in line with gasoline. The retail price was now to be decided by the market forces and GOI will no longer have to compensate OMCs for selling diesel below market prices. This step was taken to incentivize firms engaged in biodiesel production in India.
CY, 2015	On January 16, the Union Cabinet chaired by the Prime Minister, Shri Narendra Modi, gave its approval for amending the Motor Spirit and High Speed Diesel Control Order for Regulation of Supply, Distribution and Prevention of

(*Continued*)

(Cont.)

Date	Action
	Malpractices dated 19.12.2005. The cabinet also decided to suitably amend Para 5.11 and 5.12 of the National biofuel policy for facilitating consumers of diesel in procuring biodiesel directly from private biodiesel manufacturers, their authorized dealers and JVs of OMCs authorized by the MoPNG. This decision was taken to encourage the production and use of biodiesel in the country. Early August, OMCs launched the country's first tender for biodiesel, seeking to procure 850 million liters from local producers between August 2015 and March 2016. As the price of diesel was already deregulated, private biodiesel manufacturers were encouraged to sell biodiesel directly to consumers subject to their product met the prescribed BIS standards. On August 10, GOI had issued notification to allow the sale of biodiesel (B100) by private manufacturers in bulk. Bids were invited until August 19. The policy was meant to help with local price discovery ahead of a potential 20% blend for biodiesel in 2017. A 20% blend for ethanol was proposed but was unlikely since the current 5% blend was yet to be reached. On August 11, 2015, Minister of State (I/c), Petroleum and Natural Gas, launched sale of B5 Diesel on World Bio Fuel Day.
CY, 2016	January 21, 2016, tender was invited for the procurement of 16 million liters of biodiesel from indigenous manufacturers meeting IS 15607:2005 specification biodiesel (B100) for supply from February 2016 to March 2016. Few bulk users such as road transport companies, state transport corporations and railways depot shad utilized biodiesel for transporting goods and people until then. According to media reports, some 13.2 million liters of B100 was procured by OMC's until June 2016 and another 40 million liters was targeted to be procured until September 2016.
March, 2017	The Cabinet Committee on Economic Affairs approved closure/winding up of CREDA HPCL Biofuel Ltd (CHBL) and Indian Oil – Chhattisgarh Renewable Energy Development Agency (CREDA) Biofuels Limited (ICBL) due to various constraints such as very poor seed yield, limited availability of wasteland, high plantation maintenance cost and so on the project became unviable and Jatropha plantation activities were discontinued. Per media reports, on March 18, 2017, OMCs floated a tender for procurement of around 267.3 million liters of biodiesel (B100) for the period from April 2017 to Oct 2017. Biodiesel development in India is still in the nascent stage and oil marketing companies are getting hit hard with a new 6% excise duty on biodiesel along with a slew of other taxes that they say will make the mandated biodiesel blending unviable. Biodiesel was excise-duty-free for 10 years. The mandate's implementation has been delayed to at least May 1 from the planned April 1 start date, however, without excise duty, biodiesel is cheaper than fossil diesel but with the duty it may be more expensive. Biodiesel prices are set based on the Malaysian palm oil board's established price.

The NBM identified Jatropha (Jatrophacurcas) as the most suitable inedible oilseed for biodiesel production to help achieve a proposed biodiesel blend of 20% with conventional diesel by 2017. That target was unmet due to a host of agronomical and economic constraints. To help fill the gap, several existing biodiesel units shifted operations to adopt multiple feedstock technology, which utilizes UCO, other unusable oil fractions, animal fats, and inedible oils; this achieved a minimal (0.001) blend rate.

The central government proposed that it and several state governments will promote planting of Jatropha and other inedible oilseeds by providing fiscal incentives to various public, private, and cooperative sectors. The former Planning Commission of India (now National Institution for Transforming India (NITI) Commission) had even set an ambitious target of planting 11.2 to 13.4 million hectares of Jatropha by the end of April 2012, but fell short due to reasons mentioned above.

The market for biodiesel (B100) is nascent and will continue to grow if there is a strong commercially viable strategy for building a sustainable biodiesel industry. Growth is encouraged by deregulating diesel prices, bulk sale of biodiesel (B100) by authorized dealers, and authorization of joint ventures of OMCs and private manufacturers to supply to bulk consumers.

References

Basavaraj, G., Rao, P. P., Reddy, R., Kumar, A. A., Rao, P. S., & Reddy, B. V. S., *A Review of the National Biofuel Policy in India: A Critique of the Need to Promote Alternative Feedstocks* (2012), International Crops Research Institute for the Semi- Arid Tropics, Andra Pradesh, India.

Blanchard, R., Bhattacharya, S. C., Chowdhury, M., Chowdhury, B., Biswas, K., & Choudhury, B. K., *A review of biofuels in India: Challenges and opportunities*. Presented at: World Energy Engineering Congress (2015), Orlando, Florida, USA.

Chauhan, R. D., Sharma, M. P., Saini, R. P., & Singhal, S. K., *Biodiesel from Jatropha as transport fuel - A case study of UP state India* (2007), Journal of Scientific & Industrial Research, 66: 394–398.

Das, S., Priess, J. A., & Schweitzer, C., *Modelling regional scale biofuel scenarios - A case study for India* (2012), GCB Bioenergy, 4: 176–192.

Department of Biotechnology, Ministry of Science and Technology, *Mission Innovation - Accelerating the Clean Energy Revolution* (2017), Government of India, New Delhi, India.

Dutta, A., & Mandal, B. K., *Use of Jatropha biodiesel as a future sustainable fuel* (2014), Energy technology and Policy, 1(1): 8–14.

Jain, S., & Sharma, M. P., *Prospects of biodiesel from Jatropha in India: A review* (2018), Renewable and Sustainable Energy Reviews, 14(2): 763–771.

Lang, A., & Hazir, H. F. A., *Jatropha Oil Production for Biodiesel and Other Products a Study of Issues Involved in Production at Large Scale* (2013), World Bioenergy Association, Sudan.

Ministry of Petroleum and Natural Gas, *National Policy on Biofuels - 2018*, The Gazette of India: Extraordinary (2018), Government of India, New Delhi, India.

Shinoj, P., Raju, S. S., Chand, R., Kumar, P., & Msangi, S., *Biofuels in India: Future Challenges* (2011), National Centre for Agricultural Economics and Policy Research, New Delhi, India.

4

Biodiesel Technology

Hamid Omidvarborna and Dong-Shik Kim

4.1 Current Status of Technology

This chapter investigates the development of technology for biodiesel with a focus on biodiesel research and development (R&D) in the past decade. In-depth study and assessment are presented on major technology development, advantages/disadvantages, commercialization, and market share. Presently, biodiesel is a common type of biofuels in the world, in which its development in different countries are shown in Table 4.1.

Various technical and socioeconomic aspects are involved in the production of biodiesel. Three aspects are selected as they are regarded to play major roles in determining the price of biodiesel, which include: (a) feedstock of biodiesel, (b) extraction, and (c) production methods.

4.1.1 Biodiesel Feedstock

Globally, there are more than 350 oil-bearing crops identified as potential sources for biodiesel production. In general, biodiesel feedstock can be divided into four main categories; edible oil, non-edible oil, animal fat, and other sources. Table 4.2 shows different types of feedstock in each category.

The type and availability of feedstock (or raw materials), production method, additives used and operational costs are contributed to the cost of the biodiesel production. Of these factors, a type of feedstock contributes to the major portion in the biodiesel production cost. The availability of feedstock for producing biodiesel depends on the regional climate, geographical locations, local soil conditions, municipal and rural life styles, industrial structure, and agricultural practices of the country. Atabani et al. (2012) [10] found from the literature that feedstock alone represents 75% of the overall biodiesel production cost (Figure 4.1).

Edible oil resources such as soybean, palm oil, sunflower seed, safflower, rapeseed, coconut and peanut are considered as the first generation of biodiesel feedstock because they were the first crops to be used for biodiesel production. Their plantations have been well established in many countries around the world such as the US, Germany, and Malaysia.

TABLE 4.1

Key milestones in the development of the biodiesel industry in the world.

Date	Development
August 10, 1893	Rudolf Diesel's prime diesel engine model (fueled by peanut oil) ran on its own power for the first time in Augsburg, Germany.
1900	Rudolf Diesel showed his engine running on 100% peanut oil at the world exhibition in Paris.
August 31, 1937	A Belgian scientist, G. Chavanne, was granted a patent (422,877) entitled as "Procedure for the transformation of vegetable oils for their uses as fuels". The concept of the patent, which is known as "biodiesel" today was proposed for the first time.
1977	A Brazilian scientist, Expedito Parente, applied for the first patent of the industrial process for biodiesel.
1979	Research was initiated in South Africa. The use of transesterified sunflower oil and refining it to diesel fuel standards was carried out.
1983	The process for producing fuel-quality/engine-tested biodiesel was completed and published internationally.
November 1987	An Austrian company named as Gaskoks established the first biodiesel pilot plant.
April 1989	The Austrian company, Gaskoks, established the first industrial-scale plant.
1991	Austria's first biodiesel standard was issued in 1991.
1997	A German standard, DIN 51606, was formalized.
2002	ASTM D6751 was first published.
October 2003	A new Europe-wide biodiesel standard, DIN EN14214, was released.
September 2005	Minnesota became the first US state to mandate that all diesel fuel sold in the state blend with biodiesel, requiring a content of at least 2% biodiesel.
October 2008	ASTM published new Biodiesel Blend Specifications Standards.
November 2008	The current version of the European Standard EN 14214 was published and supersedes EN 14214:2003.
2010	Glycerine-based biofuel: A renewable fuels provider is reported to have filed a patent on a technology that allows using glycerine as main feedstock in the formulation of a stable biofuel. Inter alia, glycerine is generated as a by-product when vegetable oils or animal fats are transformed into biodiesel via transesterification. A member of the mustard family, camelina produces oil that seems to show promise as an aviation fuel, the specifications of which are distinct from biodiesel used for land-based vehicles. Extensive trials are being carried out, especially in North America and Europe, by both, military and civil aviation. In the US, for instance, half of the continental military jet fuel requirement is mandated to be met by alternative fuels by 2016. The interest of commercial airlines for renewable fuels that can be sustainably produced is also rising. As an aviation fuel, camelina oil can be blended with conventional fuel at inclusion rates as high as 50%. With final approval and certification of the fuel expected next year, private sources anticipate camelina jet fuel production to rise to over 3 million tons by 2025. Allegedly, camelina can be grown in dry regions, not best suited for food crop production. To date, the crop is cultivated primarily in

(Continued)

TABLE 4.1 (Cont.)

Date	Development
	the US and Canada, mainly in rotation with wheat and using conventional equipment. Yield improvements via varietal selection still present major challenges. Camelina is expected to remain a regionally based feedstock and will have to compete with other feedstock. A group of USA researchers reported to have simplified the conversion of used vegetable oil into biodiesel. The new conversion process is claimed to be environmentally benign and considerably faster than the conventional method. The new technology's viability at industrial scale still has to be proven.
2011	A study prepared by a respected academic institution and funded by Boeing sees jatropha-based biodiesel as a potential substitute for traditional jet fuel. The report states that jatropha can deliver strong environmental and socioeconomic benefits as it is normally grown on land unsuitable for farming. According to the study, jatropha fuel allows greenhouse gas (GHG) emission reductions of up to 60% when compared petroleum-based jet fuel. Faced with constant rises in fuel prices and calls to become more environmentally responsible, civilian as well as military aviation have shown increased interest in jatropha fuel. The United Arab Emirates government has entered a partnership with McDonald's Restaurants to produce biodiesel based on recycled cooking oil. The country's first licensed biodiesel plant will have an annual output capacity of one million liters. Used oil will be converted into 100% biodiesel on a one-to-one basis. McDonald's committed to using the biodiesel in its logistic fleet to achieve a reduction in its carbon footprint. Reportedly, in India, the private sector has launched a crop development program involving molecular breeding and biotechnology to produce high-yielding varieties adapted to the country's specific agronomic conditions. Eventually, the initiative envisages the deployment of 35,000 ha with high-performing jatropha hybrids. Since 2009, The Indian government called for the use of wasteland to cultivate non-edible oil crops such as jatropha. It has set biodiesel procurement prices, which, however, are reported not to be remunerative enough for the industry to grow. No form of financial assistance is provided to the growers or processors.
2012	Additional public funding of research on camelina oil as biodiesel feedstock has been made available in the US. The oilseed is believed to hold significant potential as a bioenergy crop in the Delta region. Planned research and outreach activities will focus on: (i) developing genetically transformed varieties suitable for the region, featuring high yield potential and other desirable traits; and (ii) analyzing the seeds oil/meal and optimizing processing technologies.
2013	McDonald's Corporation, the global foodservice retail company, reported new efforts in Australia and the United Arab Emirates to recycle vegetable oil. Waste oils are collected from the company's outlets, filtered and converted into biodiesel that is used to run the corporate truck fleet, thereby contributing to cut GHG emissions. Grown in subtropical and tropical regions, jatropha curcas has gained increasing attention for the use of its non-edible oil as a feedstock for biodiesel production. Reportedly, in the US, the food industry has been officially notified of hazardous toxins contained in parts of the plant. Found primarily in glycerin and protein by-products, the toxic compounds can also be detected in the

(*Continued*)

TABLE 4.1 (Cont.)

Date	Development
	plant's oil. Although unaware of any intentional substitution or contamination in food products from the jatropha plant, US authorities invited suppliers and manufacturers to take appropriate steps to prevent the use of ingredients that might be intentionally or otherwise adulterated with jatropha. The US Environmental Protection Agency (EPA) approved camelina oil as a biodiesel feedstock. According to EPA's evaluation, camelina oil-based diesel meets the 50% GHG reduction threshold required to qualify as "biomass-based diesel" or "advanced fuel" under the US bioenergy policy. Possible uses include transportation fuel, jet fuel as well as heating oil. In recent years, camelina oil has been extensively studied by the US military as a biofuel blendstock. Adaptable to marginal agroecological conditions, in recent years jatropha curcas attracted growing attention worldwide as a potential biofuel feedstock. However, the circumstance that only wild populations with uncertain yield performance levels were available for seed purposes hampered the crop's commercial take-off. A US energy crop company has confirmed jatropha's significant genetic diversity, which is said to make the plant well-suited for major performance gains. The employment of modern molecular and genetic techniques is expected to allow the creation of elite hybrid seeds – characterized by higher yields, improved plant health, and increased stress tolerance – within a relatively short period of time.
2014	European researchers are developing low-energy water-free methods to produce biodiesel from either virgin vegetable oil or waste cooking oils/animal fats and other fatty wastes – a technology that could offer significant environmental and economic benefits. Traditional methods use high volumes of water to remove impurities, producing large amounts of polluted wastewater. Reportedly, the proposed alternative technology relies on catalysts to pretreat biodiesel, using environment-friendly techniques for the removal of impurities. A US biotech company, SGB Inc. (SGB, 2014), informed that it has successfully created hybrid jatropha characterized by both high yield and rapid time to maturity. The new variety is said to reach maturity in 1–2 years. Reportedly, top yield achieved in trials amount to 2.6 tonnes per hectare in the first year. The company informed that the technology is ready for full-scale commercial deployment of jatropha. Commercial partnerships are envisaged in Central America, India, and Southeast Asia. The nonedible oil obtained from jatropha seed is mostly used as feedstock for biodiesel production.
2015	UK researchers devised a technology to increase productivity in biodiesel production by recycling a by-product. Reportedly, crude glycerol – a by-product of the vegetable-oil-to-biodiesel conversion process that employs methanol derived from fossil fuel – can be turned back into methanol, which can then be used as a starting reactant to create more biodiesel. Apart from lowering production costs, the technology is said to offer significant environmental benefits by sustainably improving the yield of biodiesel in a way that doesn't require additional fossil fuels. A private research institute in India developed a process for standardizing the production of coconut methyl ester (CME) from coconut oil. The research comprised optimization of CME production, studies of its physicochemical

(Continued)

TABLE 4.1 (Cont.)

Date	Development
	properties, and testing its efficiency as a fuel in diesel automobile engines. Reportedly, concerned departments at central government level have expressed interest to fund further research in this field. New research conducted in Australia seems to suggest that exhaust from rapeseed oil-based biodiesel may be more harmful to human epithelial cells than exhaust from traditional diesel. Allegedly, the ultrafine size of fuel exhaust particles from refined and blended rapeseed oil may lead to respiratory problems. In the Islamic Republic of Iran, researchers developed a process to produce biodiesel without using fossil-based methanol. The scientists replaced methanol, a key component in standard transesterification reactions, with bioethanol obtained from plant residues, thus using exclusively renewable products. In the research trials, a single raw material – the castor plant – was used to obtain both vegetable oil (extracted from the seeds) and ethanol – obtained via a biorefinery process from plant residues, mainly stems, leaves and seedcake.
2016	An Indian company specialized in bioenergy crop R&D announced the launch of a new high-yielding, pest-tolerant variety of moringa oleifera. Moringa is a fast-growing, drought-resistant tree native to northwestern India that is widely cultivated in tropical and subtropical areas for various food and non-food purposes. While to date moringa oil has been used mainly as a high-value food supplement and ingredient in cosmetic/medicinal products, the high-yielding variety developed in India is meant for use as a biodiesel feedstock.
2017	Brazilian scientists developed a new method for verifying the purity and quality of biodiesel. The detection of raw vegetable oils in biodiesel – as opposed to refined, trans-esterified oil – is said to be crucial as the presence of such oils lowers the quality of the fuel, increases pollution and can cause damage to engines. The newly developed detection method uses nuclear magnetic resonance spectroscopy.

Currently, more than 95% of the world biodiesel is produced from edible oils such as rapeseed (84%), sunflower oil (13%), palm oil (1%), soybean oil and others (2%). Table 4.3 shows primary biodiesel feedstock for some countries around the world.

Biodiesel has become more attractive recently because of its environmental benefits and the fact that it is made from renewable resources. Biodiesel can be produced from straight vegetable oils (SVOs), oils extracted from various plant species, fish and animal fats, and waste oils. Amongst many resources, availability of fat and oil sources and cost economy are the major factors that determine the large scale production of the biodiesels. There are two aspects of the cost of biodiesel, the costs of raw material (fats and oils) and the cost of processing. As mentioned above, the cost of raw materials accounts for 60 to 75% of the total cost of biodiesel fuel. It was reported that the application of used cooking oil can lower the cost significantly (Canackci

TABLE 4.2
Main feedstock for biodiesel.

Feedstocks	Types
Edible Oils	Barley, Canola, Castor, Coconut, Corn, Cottonseed, Groundnut, Linseed, Olive, Palm and palm kernel (Elaeis guineensis), Peanut, Rapeseed (Brassica napus L.), Rice bran oil (Oryza sativum), Sesame (Sesamum indicum L.), Sorghum, Soybean (Glycine max), Sunflower (Helianthus annuus), Wheat
Non-Edible Oils	Abutilon muticum, Aleurites moluccana, Babassue Tree, Camelina (Camelina Sativa), Coffee ground (Coffea arabica), Cottonseed (Gossypium hirsutum), Croton megalocarpus, Cumaru, Cynara cardunculus, Honge, Jatropha curcas, Jojoba (Simmondsia chinensis), Karanja or honge (Pongamia pinnata), Mahua (Madhuca indica), Moringa (Moringa oleifera), Nagchampa (Calophyllum inophyllum), Neem (Azadirachta indica), Pachira glabra, Passion seed (Passiflora edulis), polanga (Calophyllum inophyllum), Pongamia (Pongamia pinnata), rubber (Ficus elastica), Rubber seed tree (Hevca brasiliensis), Salmon oil, Tall (Carnegiea gigantean), Terminalia belerica, Tobacco seed, Nahor (Mesua ferrea), Palm (Erythea salvadorensis), Kokum (Garcinia indica)
Animal fats	Beef tallow, Chicken fat, Fish oil, Pork lard, Poultry fat
Other Sources	Algae (Cyanobacteria), Bacteria, Fungi, Latexes, Miscanthus, Poplar, Switchgrass, Tarpenes, Microalgae: Chlorellavulgaris, Pseudochoricystis, Phaeodactylum, Dunaliella tertiolecta, Volvox, Botryococcus braunii, Chlamydomonas, Pseudochlorococcum

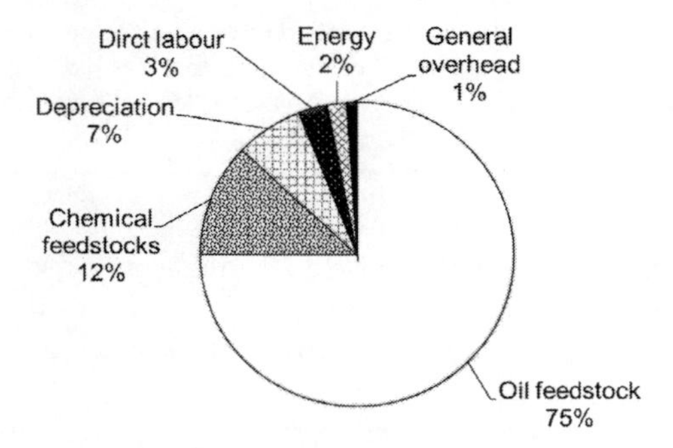

FIGURE 4.1
General cost breakdown for biodiesel production (Atabani et al., 2012; Lin et al., 2011).

and Gerpen, 2003 [11]; Zhang et al., 2003 [12]). Table 4.4 shows a list of suitable biodiesel feedstocks categorized in four titles: Top ten oilseeds, other oil crops, new varieties, and animal fats and used oils.

New varieties include modified plant oils. Modifications are carried out using hydrogenation or conventional breeding methods.

TABLE 4.3

Current biodiesel feedstock worldwide.

Country	Feedstock
US	Soybean/waste oil/peanut
Canada	Rapeseed/animal fat/soybean/yellow grease and tallow/mustard/flax
Mexico	Animal fat/waste oil
Germany	Rapeseed
Italy	Rapeseed/sunflower
France	Rapeseed/sunflower
Spain	Linseed oil/sunflower
Greece	Cottonseed
UK	Rapeseed/waste cooking oil
Sweden	Rapeseed
Ireland	Frying oil/animal fats
India	Jatropha/Pongamia pinnata (karanja)/soybean/rapeseed/sunflower/peanut
Malaysia	Palm oil/jatropha
Indonesia	Palm oil/jatropha/coconut
Singapore	Palm oil
Philippines	Coconut/jatropha
Thailand	Palm/jatropha/coconut
China	Jatropha/waste cooking oil/rapeseed
Brazil	Soybean/palm oil/castor/cotton oil
Argentina	Soybean
South Korea	Soybean/waste cooking oil
Japan	Waste cooking oil
New Zealand	Waste cooking oil/tallow

(Source: Atabani et al., 2012).

TABLE 4.4

Selection of suitable biodiesel feedstock.

Top ten oilseeds	Other oil crops	New varieties	Animal fats and used oils
Soybean	Camelina	High oleic sunflower	Tallow
Cottonseed	Hemp	High oleic rapeseed	Lard
Groundnut	Olive	Low linoleic rapeseed	Poultry fats
Sunflower	Jatropha	High erucic acid rapeseed	Rendered fats
Rapeseed	Corn		Used frying oil
Sesame			Used motor oil
Oil palm			
Coconut			
Linseed			
Castor			

(Source: Salvi and Panwar, 2012) [13]

4.1.2 Production Technology

Biodiesel is a chemically modified alternative fuel for use in diesel engines. Biodiesel was first popularized in the US by the Soy Diesel Development Board, now it is defined by the National Biodiesel Board via ASTM D6751 (EN 14214 in the EU). Table 4.5 shows the biodiesel properties required by ASTM D6751.

There are four primary ways to make biodiesel: direct use and blending, microemulsions, thermal cracking (pyrolysis), and transesterification. Transesterification is the most common method in biodiesel production. Each method is briefly reviewed, and then the notable technical progress is discussed.

4.1.2.1 Direct Use and Blending

As the name indicates, this method directly uses vegetable oil or mixes it with petroleum-based diesel. Use of vegetable oil has advantages such as liquid nature-portability, high heat content (80% of diesel fuel), availability and renewability Ma F, Hanna MA., Biodiesel production: a review. Bioresource technology, 70(1):1–15 (1999). However, vegetable oil as a substitute for diesel is not practical due to its disadvantages: (1) higher viscosity, (2) lower volatility and (3) the reactivity of unsaturated hydrocarbon chains (Pryde, 1983) [14]. In addition to that, problems occur when an engine operates on vegetable oils for long periods of time, especially with direct-injection

TABLE 4.5
ASTM D 6751–02 requirements.

Property	Method	Limits	Units
Flash point, closed cup	D 93	130 min	°C
Water and sediment	D 2709	0.050 max	% volume
Kinematic viscosity, 40 ° C	D 445	1.9–6.0	mm^2/s
Sulfated ash	D 874	0.020 max	wt. %
Total Sulfur	D 5453	0.05 max	wt. %
Copper strip corrosion	D 130	No. 3 max	
Cetane number	D 613	47 min	
Cloud point	D 2500	Report to customer	°C
Carbon residue	D 4530	0.050 max	wt. %
Acid number	D 664	0.80 max	mg KOH/g
Free glycerin	D 6584	0.02	wt. %
Total glycerin	D 6584	0.24	wt. %
Phosphorus	D 4951	0.001	wt. %
Vacuum distillation end point	D 1160	360 °C max, at 90% distilled	°C
Storage stability	To be determined	To be determined	To be determined

engines. The problems include (1) coking and trumpet formation on the injectors to such an extent that fuel atomization does not occur properly or is even prevented as a result of plugged orifices, (2) carbon deposits, (3) oil ring sticking and (4) thickening and gelling of the lubricating oil due to contamination by the vegetable oils. Blending of vegetable oil with diesel may reduce some of these problems. For example, a mixture of degummed soybean oil and No. 2 diesel in the ratio of 1:2 was reported to have no lubricating oil thickening and gelling problems.

In general, though, the direct use or blending of vegetable oil as a fuel cannot avoid the problem of oil deterioration and incomplete combustion. Polyunsaturated fatty acids in vegetable oil are very susceptible to polymerization and gum formation caused by oxidation during storage or by complex oxidative and thermal polymerization at the higher temperature and pressure of combustion. The gum did not combust completely, resulting in carbon deposits and lubricating oil thickening. In addition, high viscosity, acid composition, and free fatty acid (FFA) content are also limiting factors that do not satisfy the standard of fuel.

4.1.2.2 Microemulsion

The high viscosity of vegetable oils can be reduced by means of microemulsions or solutions made of alcohols, such as methanol or ethanol. The microscopic droplets of oil are suspended in clear or translucent, stable low-viscosity solutions. However, practical experiences and laboratory tests for motors operating on fuels created by microemulsion showed carbon deposits, injector needle sticking and an increase in the viscosity of the lubricating oil.

To solve the problem of the high viscosity of vegetable oils, microemulsions with solvents such as methanol, ethanol and 1-butanol have been studied. A microemulsion is defined as a colloidal equilibrium dispersion of optically isotropic fluid microstructures with dimensions generally in the 1–150 nm range formed spontaneously from two normally immiscible liquids and one or more ionic or non-ionic amphiphiles. They can improve spray characteristics by explosive vaporization of the low boiling constituents in the micelles (Pryde, 1984) [15].

Short term performance of both ionic and non-ionic microemulsions of aqueous ethanol in soybean oil was nearly as good as that of No. 2 diesel, in spite of the lower cetane number and energy content. Various types of emulsions were tested and they were proved to be effective in reducing viscosity and enhancing cetane number (Goering, 1984) [16].

4.1.2.3 Thermal Cracking (Pyrolysis)

This is one of the basic technologies for the production of biodiesel. Pyrolysis is a thermal decomposition under high temperatures. The process transforms original, high-viscosity triglycerides into low-viscosity combustibles with a high octane number. However, biodiesel produced by pyrolysis alone is unacceptable for many applications because of ash and carbon residues. During this process, oxygen is removed and therefore, full benefits associated with the use of biodiesel cannot be achieved.

Pyrolysis, strictly defined, is the conversion of one substance into another by means of heat or by heat with the aid of a catalyst. It involves heating in the absence of air or oxygen and cleavage of chemical bonds to yield small molecules. Pyrolytic chemistry is difficult to characterize because of the variety of reaction paths and the variety of reaction products. The first pyrolysis of vegetable oil was conducted in an attempt to synthesize petroleum from vegetable oils. Since World War I, many investigators have studied the pyrolysis of vegetable oils to obtain products suitable for fuel. In 1947, a large scale of thermal cracking of Tung oil calcium soaps was reported (Chang and Wan, 1947) [17]. Tung oil was first saponified with lime and then thermally cracked to yield a crude oil, which was refined to produce diesel fuel and small amounts of gasoline and kerosene. 68 kg of the soap from the saponification of Tung oil produced 50 L of crude oil.

In addition, catalytic cracking of vegetable oils to produce biofuels has been studied. Copra oil and palm oil stearin were cracked over a standard petroleum catalyst SiO_2/Al_2O_3 at 450 °C to produce gases, liquids and solids with lower molecular weights. The condensed organic phase was fractionated to produce biogasoline and biodiesel fuels. The chemical compositions (heavy hydrocarbons) of the diesel fractions were similar to fossil fuels. The process was simple and effective compared with other cracking processes. There was no wastewater or air pollution.

The problems that make pyrolysis unattractive are expensive equipment and process costs. Although the products are chemically similar to petroleum-derived gasoline and diesel fuel, the removal of oxygen during the thermal processing also removes any environmental benefits of using an oxygenated fuel. It produced some low-value materials and, sometimes, more gasoline than diesel fuel.

4.1.2.4 Transesterification

Transesterification is the most common method that synthesizes biodiesel from various types of feedstock. Glycerol, a highly viscous liquid, is a central component in the structure of many lipids found in animal and vegetable oils. The transesterification process inserts an alcohol, such as methanol, ethanol, propanol or butanol, into the molecular structure of those oils, allowing for the extraction of the glycerol that settles down at the bottom of the vessel. The process requires several hours, but the reaction is accelerated with the application of small amounts of liquid catalyst. For example, a mix of soybean oil and butanol in the presence of one percent sulfuric acid takes three hours to transform into biodiesel when heated to 117 °C.

Fats and oils have quite big molecules with a spinal of glycerol on which are bond three fatty acid, namely triglyceride. By the transesterification, the fatty acids are removed from the glycerol and each is bond with methanol. So, the reaction includes one mole of glycerol and three moles of fatty acid methyl ester (FAME). The liquid catalyst may be acid, base, or enzyme. The issues in the base (i.e., alkali) catalyzed transesterification process are related to chemical properties of feedstock and the reaction mechanism. The feedstock includes FFAs and water, and FFAs in the oil react with alkaline catalyst to form soaps. It results in loss of catalyst and reduction in yield. Water deactivates the catalysts and hydrolyses fats to form FFAs also. In addition, drying of oil is required.

Therefore, even if alkali-catalyzed transesterification is much faster than acid-catalyzed, when glyceride has a higher FFA content and more water, acid-catalyzed transesterification is suitable. The acids could be sulfuric acid, phosphoric acid, hydrochloric acid or organic sulfonic acid. Alkalis include sodium hydroxide, sodium methoxide, potassium hydroxide, potassium methoxide, sodium amide, sodium hydride, potassium amide, and potassium hydride (Sprules and Price, 1950) [18]. Because of the reaction rate, production yields, and costs may vary depending on the types of catalyst and feedstock, the selection should be carefully determined. Table 4.6 shows the characteristics of different biodiesel production methods.

The basic reaction scheme in biodiesel production on the 100 lbs basis is as follows:

$$100\ lbs\ of\ oil + 10\ lbs\ of\ methanol \rightarrow 100\ lbs\ of\ biodiesel + 10\ lbs\ of\ glycerol$$

The transesterification process removes the glycerides and combines oil esters of vegetable oil with alcohol (methanol or ethanol) to produce fatty acid alkyl esters (FAAEs). This process reduces the viscosity to be comparable with diesel and improves combustion. The reaction scheme is shown in Figure 4.2.

The most commonly used method for commercial production of biodiesel fuel is transesterification of vegetable oils and animal fats with alcohol. Methanol or ethanol is commonly used alcohols. The use of catalyst affects the biodiesel production in different aspects such as performance, combustion, and emission characteristics. The role of catalyst is to split the oil into glycerine and biodiesel. Most of the produced biodiesel today is completed with the base catalyzed reaction (typically sodium hydroxide or potassium hydroxide) which is dissolved in methyl alcohol (Basha and Gopal, 2012) [20]. Base-catalyzed transesterification reacts lipids (fats and oils) with alcohol catalyst to produce biodiesel and an impure co-product, glycerol as illustrated in Figure 4.2. However, if the feedstock oil is used or has

TABLE 4.6
Comparison of the different technologies to produce biodiesel.

Variable	Base	Enzyme	Supercritical	Monolithic	Resin	Acid
Temp. (°C)	60–70	30–50	200–350	50–180	60–180	50–80
Products from FFA	Soaps	Esters	Esters	Esters	Esters	Esters
Effect of water [a]	↓	↓			↓	
Yield to ester	Normal	High	High	Normal	Good	Normal
Purification of glycerol	Difficult	Simple	Simple	Simple	Simple	Difficult
Reaction time [b]	1–2 h	8–70 h	4–10 min	6 h	Variable	4–70 h
Ester purification	Difficult	Simple	Simple	Simple	Simple	Difficult
Cost	Cheapest	Expensive	Expensive	Medium	Medium	Cheaper
Amount of equipment	High	Low	Low	Low	Low	High

(Source: Marchetti et al., 2007) [19].
[a] In this case, the down arrow means that water is a drawback while the line means that is not affected and the system will be able to treat a raw material with some amount of water. For the enzyme case, a down arrow has been supplied, in this case, it is important to say that it is believed that some water is required for enzyme activation, however, a lot of water will produce a deactivation of the catalyst. In the case of the resin, it could be seen a down arrow as well as a line, this is due to the fact that water has different effects over different solid catalysts. In the case of the monolithic scenario, a line has been selected due to the fact that leaching is not caused by water per se but for a non-stability of the catalyst.
[b] The reaction time set in this table is what it is most likely, however, it is important to point out that other times for the same technology could be found in the open literature.

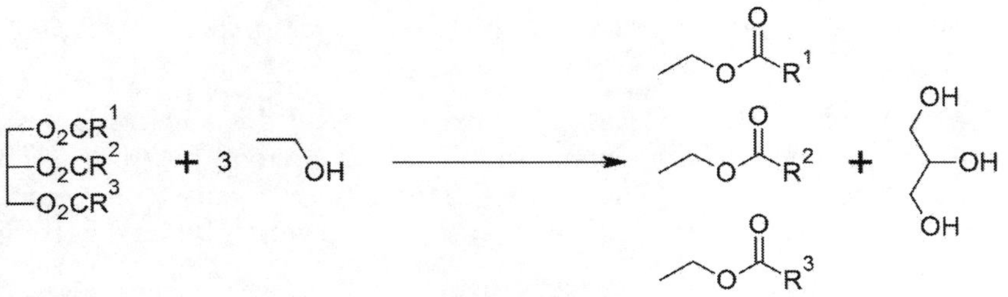

FIGURE 4.2
Transesterification scheme to produce ethyl ester fatty acid (3) and glycerol (4).

a high acid content, acid-catalyzed esterification can be used to react with fatty acids with alcohol to produce biodiesel. Other methods, such as fixed-bed reactors, supercritical reactors, and ultrasonic reactors, forgo or decrease the use of chemical catalysts.

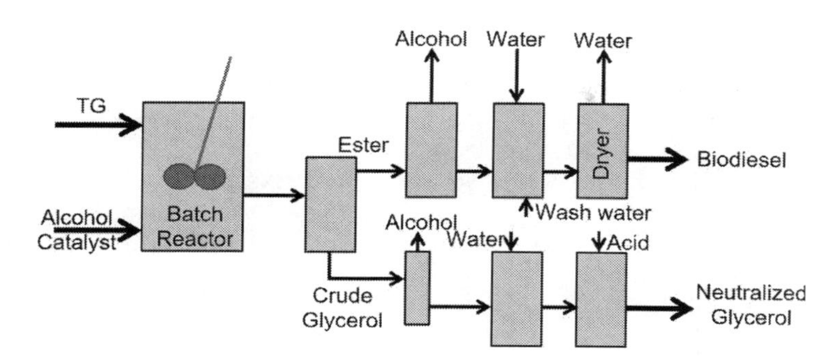

FIGURE 4.3
Flow diagram of a batch base catalyzed process of biodiesel production.

Figure 4.3 shows a flow diagram of a typical batch process for a base-catalyzed biodiesel process. Triglyceride is fed to a batch reactor with an alcohol catalyst. Ester (biodiesel) and the byproduct (glycerol) are separated. And then alcohol is separated to be reused for transesterification. Water is used to wash out the residual impurities and then is dried to produce the final product, biodiesel. In the glycerol stream, water and acid are added to neutralize glycerol. The neutralized glycerol can be used for other industrial applications.

4.1.2.5 Lipase-Catalyzed Method

Large amounts of research have focused recently on the use of enzymes as a catalyst for the transesterification. Lipase is an enzyme that catalyzes the formation or cleavage (hydrolysis) of fats (lipids). Lipases are a subclass of the esterases. Lipases perform essential roles in the digestion, transport and processing of dietary lipids (e.g., triglycerides, fats, oils) in most living organisms.

As indicated before, the high content of FFA in oils causes the problem of high viscosity and generation of impurities in the biodiesel production process. It was found that lipases make the transesterification reaction less sensitive to high FFA content. Because methanol deactivates lipase, methyl acetate (MA) is used as a replacement of methanol (lipase is not deactivated by MA) making the lipase system much more cost effective.

4.1.2.6 Pretreatment and Post-Treatment Processes

Feedstock should be pretreated in most cases before transesterification reaction. The reaction products should also be treated. The pre- and post-treatment processes vary depending on the feedstock and reaction method used. Because these processes may easily add significant extra costs and they should be carefully considered in designing the process.

Common feedstocks used in biodiesel production include yellow grease (recycled vegetable oil), "virgin" vegetable oil, and tallow (beef fat). Recycled oil should be pre-treated to remove impurities from cooking, storage, and handling, such as dirt, charred food, and water. Virgin oils are usually refined, but not to a food-grade level.

Due to the high temperature in cooking, polymerization happens for the waste cooking oil and the polymers increase the viscosity. Therefore, the degumming process to remove phospholipids and other plant matter are common. In addition to the degumming process, cooking oils may require extra pre-treatment due to the wide range of physical and chemical properties and impurities. During frying, vegetable oil undergoes various physical and chemical changes, and many undesirable compounds are formed. These compounds may cause undesirable impacts on biodiesel quality; the effect of those compounds should be well investigated. For example, when the refined rapeseed oil is heated at 180 °C for 20 h, it was observed that the polar content of oil increases because of the various physical and chemical changes (Mittelbach and Enzelsberger, 1999) [21]. After 20 h, the heated rapeseed oil was transesterified with methanol using KOH as a catalyst at room temperature. Analysis of the result was reported that, along with methyl esters, dimeric FAME was also observed. The polymers formed during the heating of oils are cleaved and during the transesterification reaction, they form monomeric and dimeric FAMEs. The oligomeric compounds formed during frying increase the molecular mass and reduce the volatility of the oil. Therefore, fatty acid esters obtained from frying oil influences the fuel characteristics (such as increasing the viscosity and reducing the burning characteristics) that lead to an increase in glycerides as well as FFAs, soaps, remaining catalyst, and other impurities. This is detrimental to the biodiesel quality. In the biodiesel ASTM standards, this value is limited to 0.05%. In the case of methyl esters obtained from heated rapeseed, it was shown that the amount of dimeric and polymeric FAMEs was increased from 0.7 wt% to 5.7 wt%.

Regardless of the feedstock, water should be removed as its presence during base-catalyzed transesterification causes the triglycerides to hydrolyze, giving salts of the fatty acids (soaps) instead of producing biodiesel. Products of the reaction include not only biodiesel, but also by-products, soap, glycerol, excess alcohol, and trace amounts of water. All of these by-products must be removed to meet the standards, but the order of removal is process-dependent. The density of glycerol is greater than that of biodiesel and this property difference is exploited to separate the bulk of the glycerol co-product. Residual methanol is typically recovered by distillation and reused. Soaps can be removed or converted into acids. Residual water is also removed from the fuel.

4.2 Recent Progress in Technology

Currently, the key problems in the production and use of biodiesel are: (1) to reduce the costs of production and (2) to avoid the competition between the production of energy and food. As a matter of fact, many vegetable oils are edible oils. Transesterification method requires a refined feedstock and therefore, the feedstock oils should be processed with an intensive pre-treatment, which occupies up to 85% of the total cost of biodiesel. Hence, feedstock for biodiesel is the area that should be improved significantly. Catalytic process is another area to be improved. The catalyst must be neutralized after the reaction, therefore, cannot be re-used and the salt formed during the neutralization contaminates the glycerol increasing the costs of its purification. In addition, more efficient and economically competent utilization methods for reaction by-product, glycerol, should be developed. An excess of glycerol is offered in a market that is already saturated.

These challenges should be overcome in order to make full use of the advantages biodiesel. Intensive R&D has been performed over the last decade, and the number of inventions published and filed showed in Figure 4.4 clearly indicates it.

A list of recent notable US patents, that address the challenges mentioned above, is given below:

1. Rautiainen M and Biogts LA, 2018. Process and system for producing liquid biofuel from bio-based oils and/or fats. US Patent 9,890,349. A process for utilizing biobased oils and/or fats for producing biofuels includes the steps of mixing alcohol with a raw material for forming a reaction mixture; pumping the reaction mixture to a reactor; mixing biogas as a catalyst

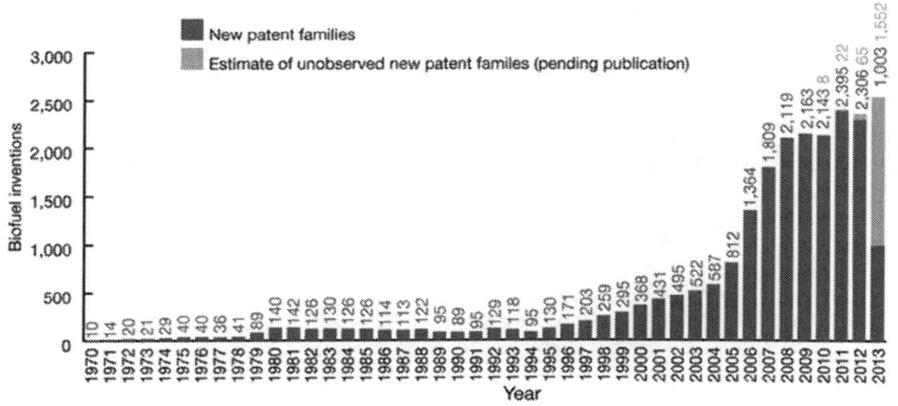

FIGURE 4.4
A number of biofuel inventions published on biodiesel.

with the reaction mixture in a selected process step either before or after the supply of the reaction mixture to a high-pressure pump, the biogas including methane and carbon dioxide; adjusting a temperature and pressure of the reactor so that the reaction mixture achieves a supercritical state; esterifying the reaction mixture to produce liquid biofuel and by-products; separating the by-products including methane and alcohol from the liquid biofuel; and recovering separated methane. An equivalent system for utilizing bio-based oils and/or fats for producing biofuels is also disclosed.

2. Hitzl M, Renz M, Ingelia SL. 2017. Biofuel product and method for the production thereof. US Patent 9,840,677.
The invention relates to a method for producing a biofuel from an aqueous mixture of carbonized biomass obtained by means of a method for the hydrothermal carbonization of biomass, characterized in that it comprises: (a) grinding the aqueous mixture of carbonized biomass until a maximum size of less than 500 micrometers of the particles contained in the mixture is obtained; (b) applying a method for the physical separation of inorganic substances; and (c) reducing the moisture content until a water content of between 25 and 55 wt. % is reached. The invention also relates to the biofuel obtained by said method, and to the use thereof in various applications.

3. Larsen TF, Andersen ER, Hjortshoj A. 2017. ORGANIC FUEL Tech AS. Process for the production of biofuel. US Patent 9,834,727.
The present invention describes a process for the production of biofuel, said process comprising, pre-treating a feedstock, mixing a catalyst with said feedstock, transferring the mixture of catalyst and feedstock into a reactor, and subjecting said mixture to a heating sequence by applying microwave energy thereto, wherein the catalyst comprises an aluminosilicate mineral, the percentage of aluminosilicate mineral in the catalyst-feedstock mixture is less than 10% (w/w), and the temperature of the mixture of catalyst and feedstock is no higher than 450 °C during the process.

4. Li L. (2010). Method of converting Triglycerides to Biofuels. US Patent 7,691,159 B2.
A triglyceride-to-fuel conversion process including the steps of (a) preconditioning unsaturated triglycerides by catalytic conjugation, cyclization, and cross-link steps; (b) contacting the modified triglycerides with hot-compressed water containing a catalyst, wherein cracking, hydrolysis, decarboxylation, dehydration, aromatization, or isomerization, or any combination thereof, of the modified triglycerides produce a crude hydrocarbon oil and an aqueous phase containing glycerol and lower molecular weight molecules, and (c) refining the crude hydrocarbon oil to produce various grades of biofuels. A triglyceride-to-fuel conversion process further including the steps of (a) carrying out anaerobic fermentation and decarboxylation/dehydration, wherein the anaerobic fermentation produces hydrogen, volatile acids, and alcohols from fermentable feedstocks, and the decarboxylation/dehydration produces alkenes from the volatile acids and alcohols, respectively; (b) feeding the alkenes to the cyclization

process; (c) feeding the hydrogen to the post refining process; and (d) recycling the aqueous phase containing glycerol to the decarboxylation/ dehydration process. A biofuel composition including straight-chain, branched and cyclo paraffins, and aromatics. The paraffins are derived from the conversion of triglycerides. The aromatics are derived from the conversion of either triglycerides, petroleum, or coal.

5. Vick B, Caspari M, Radaelli G. 2012. Aurora Algae, Inc., Hayward CA (US). Methods and compositions for production and purification of biofuel from plants and microalgae. US Patent 8,088,614 B2.
Methods and compositions are provided for producing purified oil from an organism, whether these organisms are wild type, selectively bred or genetically modified, and are suitable for the large scale production of an oil product. The organism may be an animal, a plant or a microorganism such as yeast, bacteria or algae. The organism is processed to create biomass which can be extracted to remove the lipids contained within the biomass. The extraction produces a crude extract rich in lipids and containing residual contaminants. These contaminants are removed by contacting the crude extract with a composition that comprises a nanomaterial. Subsequently, an oil product is recovered which is substantially free from residual contaminants, such as pigments.

6. Rhodes JS. 2012. Whole crop biofuel production (WCBP). US Patent 8,285,635 B2.
A computerized method of using a data processor having a memory to account for carbon flows and determine a regulatory value for a biofuel can include (i) storing, in memory, a first set of one or more carbon flow values characterizing the production and use of a biofuel derived from a first fraction of an agricultural biomass, (ii) storing, in memory, a second set of one or more carbon flow values characterizing the production and use of a co-product from a second fraction of the agricultural biomass, wherein the second fraction comprises an agricultural residue and wherein the co-product mitigates anthropogenic GHG emission, and (iii) calculating, using the data processor, a regulatory value for the biofuel from the first and second sets of carbon flow values.

7. Fan Q. 2012. Gas Technology Institute (Des Plaines, IL). Biofuel production by high temperature non-faradaic electrochemical modification of catalysis. US Patent 8,183,421 B2.
A method for producing biofuels from biomass in which a refined biomass material is introduced into a non-Faradaic electrochemical device, preferably at a temperature greater than or equal to about 150 °C, and deoxygenated and/or decarboxylated in said device to produce an increased carbon chain fuel.

8. Dumenil JC. BP Biofuels UK Ltd. Process, plant and biofuel for integrated biofuel production. US Patent 8,152,867 B2.
This invention relates to a process, a plant, and a biofuel for integrated biofuel production, such as with butanol, biodiesel, and/or sugar product.

The integrated process includes the step of removing hexose from a feedstock to form a lignocellulosic material. The process also includes the step of converting the hexose to butanol and/or a biodiesel material, and the step of depolymerizing lignocellulosic material to form pentose and a residue. The process also includes the step of converting the pentose to butanol and/or a biodiesel material.

9. Gurski SM, Kazim A, Cheung HK, Obeid A. 2012. System and process of biodiesel production. US Patent 8,192,696 B2.
A system and process for continuous production of FAME from the fatty acid triglycerides of waste oil via transesterification in the presence of a reusable sugar-based catalyst. The system and process incorporate recycling and re-use of waste by-product streams to result in near-zero emissions, with a 97% product yield mix consisting of almost pure biodiesel and a very small percentage of impurities including glycerol.

10. Yoshikuni Y, Kashiyama Y. Bio Architecture Lab, Inc. Biofuel Production. US Patent 8,211,689.
Methods, enzymes, recombinant microorganism, and microbial systems are provided for converting polysaccharides, such as those derived from biomass, into suitable monosaccharides or oligosaccharides, as well as for converting suitable monosaccharides or oligosaccharides into commodity chemicals, such as biofuels. Commodity chemicals produced by the methods described herein are also provided. Commodity chemical enriched, refinery-produced petroleum products are also provided, as well as methods for producing the same.

11. Miller SJ. 2012. Chevron U.S.A. Inc. Production of biofuels and biolubricants from a common feedstock. US Patent 8,124,572 B2.
The present invention is directed to methods and systems for processing triglyceride-containing, biologically-derived oils, wherein such processing comprises conversion of triglycerides to FFAs and the separation of these fatty acids by saturation type. Such separation by type enables the efficient preparation of both lubricants and transportation fuels from a common source using a single integrated method and/or system.

12. Berry WW, Tegen MG, Hillis SL. 2012. Inventure Chemical, Inc. Production of biodiesel, cellulosic sugars, and peptides from the simultaneous esterification and alcoholysis/hydrolysis of oil-containing materials with cellulosic and peptidic content. US Patent 8,212,062 B2.
The present invention relates to a method for producing FAAEs as well as cellulosic simplified sugars, shortened protein polymers, amino acids, or combination thereof resulting from the simultaneous esterification and hydrolysis, alcoholysis, or both of algae and other oil containing materials containing FFA, glycerides, or combination thereof as well as polysaccharides, cellulose, hemicellulose, lignocellulose, protein polymers, or combination thereof in presences of an alcohol and an acid catalyst.

13. Jackam JP, Pierce JM, Jones JD. 2012. Seneca Landlord, LLC. Production of biodiesel and glycerin from high free fatty acid feedstocks. US Patent 8,088,183 B2.
A system and method for the conversion of FFAs to glycerides and the subsequent conversion of glycerides to glycerin and biodiesel include the transesterification of a glyceride stream with an alcohol. The FAAEs are separated from the glycerin to produce a first liquid phase containing a FAAE rich (concentrated) stream and a second liquid phase containing a glycerin-rich (concentrated) stream. The FAAE rich stream is then subjected to distillation, preferably reactive distillation, wherein the stream undergoes both physical separation and chemical reaction. The FAAE rich stream is then purified to produce a purified biodiesel product and a glyceride rich residue stream. The glycerin-rich second liquid phase stream may further be purified to produce a purified glycerin product and a (second) wet alcohol stream.

14. Zhou G, Munson C, Elomari S, Roby SH. 2011. Chevron U.S.A. Inc. Enhanced biodiesel fuel having improved low-temperature properties and methods of making same. US Patent 8,057,558 B2.
The present invention is generally directed to novel biodiesel fuel compositions having enhanced low-temperature properties. The present invention is additionally directed to methods (i.e., processes) for making such enhanced biodiesel fuels by improving the low-temperature properties of ester-based biodiesel fuels via in situ enhancement and/or additive enhancement.

15. Kale A. 2012. Heliae Development, LLC. Methods of and systems for producing biofuels. US Patent 8,152,870 B2.
A method of making biofuels includes dewatering substantially intact algal cells to make an algal biomass, sequentially adding solvent sets to the algal biomass, and sequentially separating solid biomass fractions from liquid fractions to arrive at a liquid fraction comprising neutral lipids. The method also includes esterifying the neutral lipids, separating a water-miscible fraction comprising glycerin from a water immiscible fraction comprising fuel esters, carotenoids, and omega-3 fatty acids. The method also includes obtaining a C16 or shorter fuel esters fraction, a C16 or longer fuel ester fraction, and a residue comprising carotenoids and omega-3 fatty acids. The method includes hydrogenating and deoxygenating at least one of (i) the C16 or shorter fuel esters to obtain a jet fuel blend stock and (ii) the C16 or longer fuel esters to obtain a diesel blend stock.

4.2.1 Feedstock Development

Nowadays, the most employed feedstocks in biodiesel production are rapeseed, sunflower, soybean and palm oils. With the conventional technology, these oils must be highly refined for eliminating FFAs before the use. Moreover, some of these oils are commonly used also for nutrition and there is competition between the production of oils for food and/or for energy. This competition is also related to the use of land for cultivation. Some other less expensive oils not in competition with food have recently considered as possible future feedstock candidates. In particular, waste oils of any sort, oil from Jatropha curcas and from algae. Jatropha is a plant growing with high oil productivity in residual drylands; while, algae grows in water ponds with extraordinary productivity. Therefore, all these oil sources and similar alternatives are today object of very intensive studies.

As stated before, the type of feedstock determines 70–85% of total manufacturing costs. Therefore, it is imperative to develop or identify high oil-content and easy-to-process feedstock to reduce the price of biodiesel. Some oil-bearing plants produce valuable byproducts such as antibiotics or pesticides, which would overshadow the pretreatment cost.

More than 350 oil-bearing crops have been identified, of which only soybean, palm, sunflower, safflower, cottonseed, rapeseed, and peanut oils are considered potential alternative fuels for diesel engines. Beef and sheep tallow and poultry oil from animal sources and cooking oil are also sources of raw materials. There are various other biodiesel sources such as almond, andiroba (Carapa guianensis), babassu (Orbignia sp.), barley, camelina (Camelina sativa), coconut, copra, cumaru (Dipteryx odorata), Cynara cardunculus, fish oil, groundnut, Jatropha curcas, karanja (Pongamia glabra), laurel, Lesquerella fendleri, Madhuca indica, microalgae (Chlorella vulgaris), oat, piqui (Caryocar sp.), poppy seed, rice, rubber seed, sesame, sorghum, tobacco seed, and wheat.

Regionally, soybean oil is commonly used in the US and rapeseed oil is used in many European countries for biodiesel production, whereas, coconut oil and palm oils are used in Malaysia and Indonesia for biodiesel production. In India and Southeast Asia, the Jatropha tree (Jatropha cursas), Karanja (Pongamia pinnata) and Mahua (M. indica) are used as significant fuel sources. New plant oils that are under consideration include mustard seed, peanut, sunflower, and cottonseed. Specially bred mustard varieties can produce reasonably high oil yields and have the added benefit that the meal left over after the oil has been pressed out can act as an effective and biodegradable pesticide. The most commonly considered animal fats include those derived from poultry, beef, and pork.

Unequivocally the most attractive source of biodiesel would be microalgae. Algae can grow practically anywhere where there is enough sunshine. Some algae can grow in saline water. All algae contain proteins, carbohydrates, lipids and nucleic acids in varying proportions. While the percentages vary with the type of algae, there are algae types that are comprised up to 40% of their overall mass by fatty acids. The most significant distinguishing characteristic of algal oil is its yield and hence its biodiesel yield. According to some estimates, the yield (per acre) of oil from algae is over 200 times the yield from the best-performing plant/vegetable oils. Microalgae are the fastest-growing photosynthesizing organisms. They can complete an entire growth cycle every few days. Approximately 46 tons of oil/hectare/year can be produced from diatom algae. Different algae species produce different amounts of oil. Some algae produce up to 50% oil by weight. According to some estimates, the yield (per acre) of oil from algae is over 200 times the yield from the best-performing plant/vegetable oils (Sheehan et al., 1998) [22]. Approximately, 46 tons of oil per hectare per year can be produced from diatom algae. Different algae species produce different amounts of oil. Microalgae have much faster growth rates than terrestrial crops. The per unit area yield of oil from algae is estimated to be between 5000 and 20,000 gallons per acre per year, which is 7–31 times greater than the next best crop, palm oil. As of 2017, although there are several companies, for example, Solazyme, recently rebranded as TerraVia, San Francisco, USA, DSM, Netherlands, ADM, Chicago, USA, and Roquette, France, that operates commercial-scale algae production, they are mostly sold as food, cosmetics, fertilizer, and nutrients products, not biodiesel.

Arguably the most promising raw material for biodiesel in the future may be waste oils. For example, biodiesel obtained from waste cooking vegetable oils has been evaluated as a cost-effective and high-quality feedstock that requires less pre-treatment. Waste cooking oil is available with a relatively cheap price for biodiesel production in comparison with fresh vegetable oil costs. Table 4.7 compares the properties of waste cooking oil, biodiesel from waste cooking oil and commercial diesel fuel. The properties of biodiesel and diesel fuels, in general, show many similarities, and therefore, biodiesel is rated as a realistic fuel as an alternative to diesel. The conversion of waste cooking oil into methyl esters through the transesterification process approximately reduces the molecular weight to one-third, reduces the viscosity by about one-seventh, reduces the flash point slightly and increases the volatility marginally, and reduces pour point considerably.

4.2.2 Manufacturing Process Development

4.2.2.1 Supercritical Transesterification

In the conventional transesterification of animal fats and vegetable oils for biodiesel production, FFAs and water always produce negative effects since the presence of FFAs and water causes soap formation, consumes the catalyst, and reduces catalyst effectiveness, all of which results in low conversion. The transesterification reaction may be carried out using either basic or acidic catalysts, but these processes require

TABLE 4.7
Comparison of biofuel properties of waste cooking oil and commercial diesel fuel.

Fuel property	Waste cooking oil	Biodiesel from waste cooking oil	Commercial diesel fuel
Kinematic viscosity (mm^2/s, at 313 K)	36.4	5.3	1.9–4.1
Density (kg/L, at 288 K)	0.924	0.897	0.075–0.840
Flash point (K)	485	469	340–358
Pour point (K)	284	262	254–260
Cetane number	49	54	40–46
Ash content (%)	0.006	0.004	0.008–0.010
Sulfur content (%)	0.09	0.06	0.35–0.55
Caron residue (%)	0.46	0.33	0.35–0.40
Water content (%)	0.42	0.04	0.02–0.05
Higher heating value (MJ/kg)	41.40	42.65	45.62–46.48
FFA (mg KOH/g oil)	1.32	0.10	-
Saponification value	188.2	-	-
Iodine value	141.5	-	-

(Source: Demirbas, 2009).

relatively time-consuming and complicated separation of the product and the catalyst, which results in high production costs and energy consumption. To overcome these problems, Kusdiana and Saka (2001) [23] and Demirbas (2002) [24] have proposed that biodiesel fuels may be prepared from vegetable oil via noncatalytic transesterification with supercritical methanol (SCM). SCM is believed to solve the problems associated with the two-phase nature of normal methanol/oil mixtures by forming a single phase as a result of the lower value of the dielectric constant of methanol in the supercritical state. As a result, the reaction was found to be complete in a very short time.

Compared with the catalytic processes under barometric pressure, the SCM process is non-catalytic, involves a much simpler purification of products, has a lower reaction time, is more environmentally friendly, and requires lower energy use. However, the reaction requires temperatures of 525–675 K and pressures of 35–60 MPa. The stoichiometric ratio for transesterification reaction requires 3 moles of alcohol and 1 mole of triglyceride to yield 3 moles of fatty acid ester and 1 mole of glycerol. Higher molar ratios result in greater ester production in a shorter time. In one study, the vegetable oils were transesterified at 1:6 to 1:40 vegetable oil–alcohol molar ratios in catalytic and supercritical alcohol conditions (Demirbas, 2002) [24]. Figure 4.5 shows a two-stage continuous biodiesel production process with subcritical water and SCM.

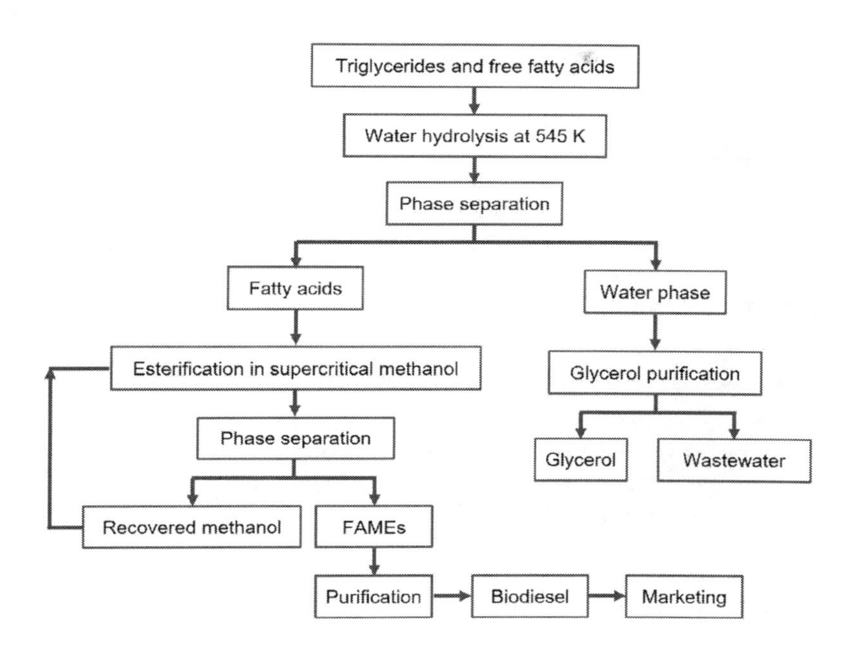

FIGURE 4.5
Flow diagram for a continuous biodiesel production process with subcritical water and SCM stages (Demirbas, 2009) [25].

4.2.2.2 Heterogeneous Catalysts

Currently, the most common method for biodiesel production is alkali-catalyzed transesterification of fatty acids in a homogeneous liquid phase. The homogeneous process needs steps of glycerol separation, washings, very stringent and extremely low limits of Na, K, glycerides, and moisture limits in biodiesel. These separation steps add high manufacturing costs raising overall production costs of biodiesel. To mitigate these costs, heterogeneous catalyzed production of biodiesel has emerged as a preferred route as it is environmentally benign needs no water washing and product separation is much easier.

The main advantage is that it requires lower investment costs than the homogeneous system, as there is no need for separation steps such as methanol/catalyst, biodiesel/catalyst and glycerine/catalyst. Recent research used CaO and CaO modified with alkaline and alkaline earth metal catalysts, and showed very good catalytic performances with high activity and stability (Gomes Joao et al., 2012) [26]. Biodiesel (FAME) yields were observed higher than 94% without expensive intermediate reactivation procedures.

One of the interesting technologies developed in recent studies is found in Badday et al. (2013) [27] Tunstohosphric acid catalyst doped on activated carbon was used to convert Jatropha oil. Ultrasound was also applied during the production process. Influences of ultrasonic energy on different process variables were elucidated. Reaction variables, that is, reaction time (10–50 min), reactants' molar ratio (5:1–25:1), ultrasonic amplitude (30–90% of the maximum sonifier power) and catalyst amount (2.5–4.5 w/w oil) were presented. A yield of 91% was achieved in just 40 min at a moderate ultrasonic amplitude (~60%), high molar ratio (25:1) and low reaction temperature (65 °C).

4.2.2.3 Reactive Extraction

Biodiesel is a fuel derived from renewable resources such as edible and inedible oil-bearing seed, algae, and waste cooking oil. The conventional biodiesel process involves oil extraction, refining and transesterification. Alternatively, transesterification can actually be performed directly from the oil-bearing materials without prior extraction. This route which is often termed "reactive extraction" or "in situ transesterification" has the advantages of simplifying the biodiesel production process as well as potentially reducing production costs. Zakaria and Harvey (2012) [28] used the reactive extraction technique with rapeseed and methanol. A high yield of ester (>85%) was achieved at high solvent to oil molar ratios (>475:1). The technique requires less processing steps than conventional techniques and is regarded as promising for rapid and cost-effective production of biodiesel.

4.2.2.4 Algae Biodiesel

Biodiesel from algae has been one of the most buzz words in renewable energy fields. However, costs for dewatering/drying, extraction, and processing have limited commercial scale production of biodiesel from algal biomass. A tremendous amount of R&D has been undergoing over the last 5 years, and the algae biodiesel R&D is expected to increase more in the future. Recent progress uses a wet lipid extraction procedure that was capable of extracting 79% of transesterifiable lipids from wet algal biomass (84% moisture) via acid and base hydrolysis at 90 °C and ambient pressures (Sathish and Sims, 2012) [29]. 76% of those extracted lipids were isolated, by further processing and converted to FAMEs. The procedure was capable of removing chlorophyll contamination of the algal lipid extract through precipitation. Another advantage of the process is its side streams that serve as feedstocks for microbial conversion to additional bioproducts. The capability of the procedure to extract lipids from wet algal biomass, reduce/remove chlorophyll contamination, potentially reduce organic solvent demand, and generate feedstocks for high-value bioproducts presents opportunities to reduce costs of scaling up algal lipid extraction for biodiesel production.

Of the numerous R&D articles published on algae biodiesel recently, a development of recombinant DNA technology appears to be notable. The idea of microalgae genetic engineering to increase their valuable compounds is very alluring. Most of genetic improvement of microalgae has been focused on the production of useful materials applicable to the cosmetic and medical fields. Due to the absence of cell differentiation, it seems that genetic manipulations of microalgae should be a much simpler system compared with higher plants. In addition, allelic genes are usually absent because of the haploid nature of most vegetative stages of microalgae. Nevertheless, the progress in the genetic engineering of microalgae was extremely slow and until last 10 years, little work has been done by adopting a genetic engineering approach to improve the microalgae.

The methodology development for microalgal transformation has advanced significantly in the last 15 years. Genetic modification and molecular tools have been developed for the green (Chlorophyta), red (Rhodophyta), and brown (Phaeophyta) algae; diatoms; euglenids; and dinoflagellates. More than 30 different strains of microalgae have been transformed successfully to date. The major problems in developing microalgae genetic engineering are the low growth rate of microalgae and also the quantity of gene expression in the microalgae species. Previous works showed that the expression efficiency of exogenous genes is mostly 0.1–0.9% of soluble proteins in host cells of microalgae, therefore, most of the new works are focused on enhancing gene expression in microalgae as heterologous hosts.

The application of genetic engineering to improve biodiesel production in eukaryotic microalgae is in its infancy, but significant advances in the development of genetic manipulation tools have recently been achieved with microalgal model systems and are being used to manipulate central carbon metabolism in these organisms. Recently, due to the availability of genome databases and the previous studies, fatty acid biosynthesis pathways were characterized. It is obvious that biosynthesis of fatty acid occurs in the plasmid of plants and microalgae before translocation to the cytoplasm for further assembly into diacylglyceride and triacylglyceride molecules. Many enzymes involved in the biosynthesis of fatty acids are encoded by single genes and are targeted to the mitochondria where fatty acid precursors are required to produce essential cofactors for mitochondrial enzyme activity. Plants fatty acids and triacylglycerides are fairly consistent, but the microalgal lipids are variable and frequently composed of triacylglycerides and polyunsaturated fatty acids that are prone to undesirable oxidation reactions affecting downstream biodiesel applications.

There are some strategies to engineer biosynthesis of fatty acids in microalgae toward more compatible lipid profiles, including secretion of lipid to from the cells to the media, over-expression of major enzymes involved in biosynthesis of fatty acids, increasing the availability of precursor molecules such as acetyl-CoA, downregulating the catabolism of fatty acids by inhibiting β-oxidation or lipase hydrolysis, altering saturation profiles through the introduction or regulation of desaturases and finally, optimization of length of fatty acid chains with thioesterases.

One of the most costly downstream processing steps in biodiesel production using microalgal feedstock is the extraction of fuel precursors from the biomass. One possible solution is to manipulate the biology of microalgal cells to allow for the secretion of lipids into the growth medium. There are in fact several pathways in nature that lead to the secretion of hydrophobic compounds, including FFAs, alkanes, and wax esters. Previous works on the secretion of FFAs in yeast have shown that inactivation of genes involved in β-oxidation and fatty acyl-CoA synthetase activity resulted in fatty acid secretion in some instances (Michinaka et al., 2003 [30]; Nojima et al., 1999) [31]. These genes were identified to have a function in fatty acid secretion in yeast. Similar screening methods could be utilized to identify microalgae that have the ability to secrete fatty acids.

4.2.2.5 Byproduct Treatment

Glycerol, one of the major by-products in the transesterification reaction, should be separated from biodiesel. The separation process should be cost-effective, and the separated glycerol should be utilized properly to increase the economic value of biodiesel. As above-mentioned, in the production of biodiesel glycerol is obtained as a by-product with a yield of 10 wt%. An excess of this substance is, therefore, available on the market and it becomes imperative to find new convenient uses for glycerol also in the perspective of reducing the cost of biodiesel production.

Despite the large number of proposals and R&D, in practice, there are only two possible strategies to consume a large amount of glycerol derived from biodiesel: (i) as feedstock for obtaining commodities. Two examples of this type are the glycerol hydrochlorination and the dehydration of glycerol to acrolein. The glycerol hydrochlorination allows obtaining chlorohydrins that are important intermediates for producing epichlorohydrin (a monomer of epoxide resins). The second example is the glycerol dehydration to acrolein for obtaining in a successive oxidation step acrylic acid. The second step is a well-known technology. Acrylic acid is produced by propene in two oxidation steps the first giving acrolein and the second acrylic acid using two different catalysts. In both described cases glycerol substitutes (ii) the use of glycerol for producing oxygenated additives for fuels.

Different substances derived from glycerol can be used as blending components for diesel fuel such as ethers (glycerol isobutyl ether), esters (triacetin), acetals and ketals. However, etherification to obtain glycerol tri-butyl ethers (GTBE) is regarded as the most promising reaction. The target product is a mixture of the di- and tri-butyl-ethers of glycerin (h-GTBE), a new additive for diesel (both fossil and bio) and also for gasoline as octane-booster. In diesel and biodiesel at 7.5 wt%, it will lead to the reduction of the emissions of particulates, NOx and unburned hydrocarbons. Moreover, blending diesel or biodiesel with h-GTBE reduces also viscosity, cloud point and pour point.

4.2.2.6 Use of Solid Waste for Biodiesel Production

The use of biodiesel is justified because of reduced carbon emissions during combustion. However, the current feedstock for the production of biodiesels can actually use more fossil fuels and emit more carbon dioxide. Use of food sources such as soybean may cause a price increase in the food market. Note that there are an estimated 3.7 billion malnourished people in the world. The use of agricultural waste and used oil from the petrochemical and restaurant industries is increasing as a result of these concerns.

Along with that trend, municipal solid waste (MSW) is getting attention for a potential source of biodiesel. MSW poses a very big challenge to local governments throughout the globe. The cities are facing constant problems in handling the MSW in their communities. MSW consists of biodegradable organic, plastic waste and paper waste. The biodegradable organics is a valuable resource for the production of biofuels and can serve an inexpensive feedstock. Recently a number of companies in North America have come up with innovative solutions to the MSW problems. For example, Terrabon, Harvest Power, Agilyx, Genomatica, Agnion, and MicreoGreen have developed technologies to convert a portion of waste into biofuels. All these companies are involved in building real-life plants which are expected to be online in 2013 in different cities in North America.

At present, cities around the globe generate about 1.3 billion tons of solid waste per year and this volume is expected to increase to 2.2 billion tons by 2025. The Medex Green Company of Canada (www.medexgreen.com) has suggested an innovative procedure to separate different components of the MSW (see Figure 4.6). BRI is conducting technology development studies for four biodiesel startup companies based in North Carolina (NC). Each company is commercializing a technology that will help grow the biodiesel industry in NC and in the US. Technology examples include:

- Converting biomass and other waste streams into biodiesel.
- Converting biodiesel waste products into high-value chemicals.

REG Houston, LLC is a 35 million gallon per year, commercial scale, and state-of-the-art biodiesel production facility in Houston, Texas. The facility was engineered and constructed by Lurgi and owned by US Biodiesel Group. In 2008, REG acquired the facility and production began the same year. REG Houston currently employs more than a dozen full-time, green-collared, skilled employees with family-wage jobs, and full benefits. Other indirect jobs include truck drivers for the hundreds of inbound and outbound trucks each month. The diagram in Figure 4.6 shows the procedure of the method for solid waste according to the present invention. The video of the procedure can be seen on their website.

Table 4.8 shows yearly projections of biodiesel production from 2009 to 2015 by the major biodiesel companies worldwide. It is clear the companies are expanding algae biodiesel significantly.

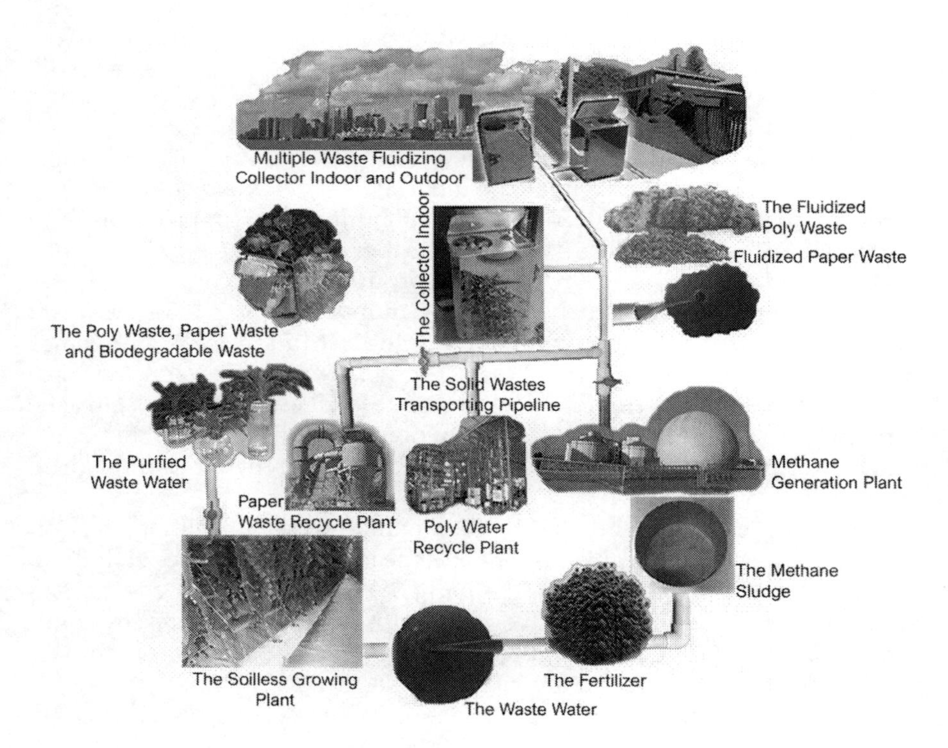

FIGURE 4.6
A suggestion for MSW utilization by The Medex Green Company.

TABLE 4.8

Biodiesel projects all over the world.

	Production Capacity (millions of gallons per year)								Project Details	
Project	09	10	11	12	13	14	15	Country	Technology	Feedstock(s) Fuel Category
Algae.Tec	0.00	0.00	0.00	0.01	0.01	0.01	0.01	Australia	Algae transesterification	Algae-Cellulosic biofuel
Aurora Algae – pilot	0.01	0.01	0.01	0.50	0.50	0.50	0.50	USA	Algae transesterification	Algae-Biomass-based diesel
Aurora Algae – demonstration	0.00	0.00	0.02	0.02	0.02	0.02	0.02	Australia	Algae transesterification	Algae-Biomass-based diesel
BARD	0.00	0.00	0.00	10.00	10.00	10.00	10.00	USA	Algae transesterification	Algae-Biomass-based diesel
BioProcess Algae	0.00	0.00	0.01	0.01	0.01	0.01	0.01	USA	Algae transesterification	Algae-Advanced biofuel
Clearfuels	0.00	0.00	0.07	0.07	0.07	0.07	0.07	USA	Steam reform/FT	Woodwaste/bagasse-Advanced Biofuel
Clearfuels	0.00	0.00	0.00	0.00	0.00	20.00	20.00	USA	Steam reform/FT	Woodwaste-Advanced Biofuel
Clearfuels	0.00	0.00	0.00	0.00	0.00	0.00	18.00	USA	Steam reform/FT	Bagasse/cane trash-Advanced Biofuel
Diamond Green	0.00	0.00	0.00	137.00	137.00	137.00	137.00	USA	Hydroprocessing	Animal residue-Advanced biofuel
Dynamic Fuels	0.00	75.00	75.00	75.00	75.00	75.00	75.00	USA	FT	Animal wastes-Advanced Biofuel
ENN	0.01	0.01	0.01	0.01	0.01	0.01	0.01		Algae transesterification	Biomass-based diesel

(Continued)

TABLE 4.8 (Cont.)

	Production Capacity (millions of gallons per year)								Project Details	
Project	09	10	11	12	13	14	15	Country	Technology	Feedstock(s) Fuel Category
Green Star Products	0.00	0.00	0.00	2.00	2.00	2.00	2.00	USA	Algae transesterification	Algae-Advanced Biofuel
Kumho Petrochemical	0.00	0.00	0.39	0.39	0.39	0.39	0.39	Korea	Algae transesterification	Algae-Biomass-based diesel
LiveFuels	0.01	0.01	0.01	0.01	0.01	0.01	0.01	USA	Algae transesterification	Algae-Biomass-based diesel
MBD Energy	0.00	0.00	0.01	0.01	3.00	3.00	3.00	Australia	Algae transesterification	Algae
Neste Oil/Singapore	0.00	223.00	223.00	223.00	223.00	223.00	223.00	Singapore	Hydroprocessing	Palm, Rapeseed oil, waste fat
Neste Oil/Finland	109.00	109.00	109.00	109.00	109.00	109.00	109.00	Finland	Hydroprocessing	Palm, Rapeseed oil, waste fat-Advanced Biofuel – palm oil
Neste Oil/Rotterdam	0.00	0.00	240.00	240.00	240.00	240.00	240.00	Netherlands	Hydroprocessing	Palm, Rapeseed oil, waste fat
Pond Biofuels	0.00	0.00	0.01	0.01	0.01	0.01	0.01	Canada	Algae burning/ transesterifcation	Algae-Biomass-based diesel
Seambiotic	0.01	0.01	0.01	0.01	0.01	0.01	0.01	Israel	Algae transesterification	Algae-Advanced Biofuel
Solix	0.01	0.01	0.01	0.01	0.01	0.01	0.01	USA	Algae transesterification	Algae-Advanced Biofuel

References

Atabani AE, Silitonga AS, Badruddin IA, Mahlia TMI, Masjuki HH, Mekhilef F., *A comprehensive review on biodiesel as an alternative energy resource and its characteristics*. Renewable and Sustainable Energy Reviews 16(4):2070–2093 (2012).

Badday AS, Abdullah AZ, Lee K-T., *Optimization of biodiesel production process from Jatropha oil using supported heteropolyacid catalyst and assisted by ultrasonic energy*. Renewable Energy: An International Journal 50:427–432 (2013).

Basha SA, Gopal KR., *A review of the effects of catalyst and additive on biodiesel production, performance, combustion and emission characteristics*. Renewable and Sustainable Energy Reviews 16:711–717 (2012).

Canackci M., Gerpen JV., *A pilot plant to produce biodiesel from high free fatty acid feedstocks*. Transactions of the ASAE 46(4):945–954 (2003).

Chang CC, Wan SW., *China's motor fuels from tung oil*. Industrial & Engineering Chemistry 39:1543–1548 (1947).

Demirbas A., *Biodiesel from vegetable oils via transesterification in supercritical methanol*. Energy Convers Manage 43:2349–2356 (2002).

Demirbas A., *Progress and recent trends in biodiesel fuels*. Energy Conversion and Management 50: 14–34. Department of energy's aquatic species program - biodiesel from algae. (2009).

Goering C.E., *Final Report for Project on Effect of Nonpetroleum Fuels on Durability of Direct-Injection Diesel Engines under Contract 59-2171-1-6-057-0*, USDA, ARS, Peoria, IL. (1984).

Gomes Joao FP, Puna Jaime FB, Bordado Joao CM, Correia M, Joana N, Dias Ana PS., *Status of biodiesel production using heterogeneous alkaline catalysts*. International Journal of Environmental Studies 69(4):635–653 (2012).

Kusdiana, D. and Saka, S., 2001. Kinetics of transesterification in rapeseed oil to biodiesel fuel as treated in supercritical methanol. *Fuel, 80*(5), pp. 693–698.

Li L., *Method of Converting Triglycerides to Biofuels* United States Patent US 7,691,159 B2, (2010).

Lin, L., Cunshan, Z., Vittayapadung, S., Xiangqian, S. and Mingdong, D., *Opportunities and challenges for biodiesel fuel*. Applied Energy 88(4):1020–1031 (2011).

Marchetti JM, Miguel VU, Errazu AF., *Possible methods for biodiesel production*. Renewable and Sustainable Energy Reviews 11:1300–1311 (2007).

Michinaka Y, Shimauchi T, Aki T, Nakajima T, Kawamoto S, Shigeta S., *Extracellular secretion of free fatty acids by disruption of a fatty acyl-CoA synthetase gene in Saccharomyces cerevisiae*. Journal of Bioscience Bioengineering 95:435–440 (2013).

Mittelbach M, Enzelsberger H., *Transesterification of heated rapeseed oil for extending diesel fuel*. Journal of the American Oil Chemists' Society 76:545–550 (1999).

Nojima Y, Kibayashi A, Matsuzaki H, Hatano T, Fukui S., *Isolation and characterization of triacylglycerol-secreting mutant strain from yeast, Saccharomyces cerevisiae*. Journal of General Applied Microbiology 45:1–6 (1999).

Pryde EH., *Vegetable oil as diesel fuel: Overview*. Journal of the American Oil Chemists' Society 60:1557–1558 (1983).

Pryde EH., *Hydroformylation of unsaturated fatty acids*. Journal of the American Oil Chemists' Society 61(2):419–425 (1984).

Salvia BL, Panwar NL., *Biodiesel resources and production technologies – A review*. Renewable and Sustainable Energy Reviews 16(6):3680–3689 (2012).

Sathish A, Sims RC., *Biodiesel from mixed culture algae via a wet lipid extraction procedure*. Bioresource Technology 118:643–647 (2012).

SGB., *Genetic Advancements and New Yield Data Drive Financing and Commercialization*. Market Wired. September 09 (2014).

Sheehan J, Dunahay T, Benemann J, Roessler P., *A Look Back at the US Department of Energy's Aquatic Species Program: Biodiesel from Algae* (NREL/TP-580-24190), Golden, CO. (1998).

Socha A, Sello J., *Efficient conversion of triacylglycerols and fatty acids to biodiesel in a microwave reactor using metal triflate catalysts*. Organic Biomolecular Chemistry 8:4753–4756 (2010).

Sprules FJ, Price D., *Production of Fatty Esters*, U.S. Patent 2,494,366, January 10, (1950).

Zakaria R, Harvey AP., *Direct production of biodiesel from rapeseed by reactive extraction/in situ transesterification*. Fuel Processing Technology 102:53–60 (2012).

Zhang Y, Dubé MA, McLean DD, Kates M., *Biodiesel production from waste cooking oil: 2. Economic assessment and sensitivity analysis*. Bioresource Technology 90 (3):229–240 (2003).

5

Production of Biodiesel from Microalgae

A Review of the Microalgae-Based Biorefinery

Mohammad-Matin Hanifzadeh and Zahra Nabati

Department of Chemical Engineering, the University of Toledo, Toledo, USA

5.1 Introduction

The commercialization of biodiesel are faced with several challenges: 1 – finding a suitable feedstock which does not compete with food resources, 2 – finding an environmental and economic sustainable strategy to produce the feedstock of the biodiesel, 3 – reduce the amount of residuals from biodiesel production by increasing the lipid content in the feedstock, 4 – finding an efficient method for conversion of feedstock to biodiesel. Between various sources of biodiesel, microalgae has recently gained much attention as it has the potential to address the aforementioned challenges since (a) it can grow in non-arable land with significantly higher productivity relative to terrestrial plant without competing with food production [1], (b) the algal production is more sustainable relative to terrestrial plants cultivation as microalgae can grow using a low quality water (saltwater, wastewater) [2] and (c) the lipid content of microalgae can reach up to 30–40% which is significantly higher comparing to other plants [3].

The overall process for production of biodiesel from microalgae is shown in Figure 5.1. Briefly, the process initiates with cultivation of microalgae in a system supplied with light, nutrient and water. After cultivation, the culture (0.7–2 g/L) is transported to a dewatering system where the microalgae cells were separated from the media. The conventional biodiesel production process is followed by the drying of wet biomass. In downstream, the dried biomass is converted to final product by addition of solvents (e.g., methanol and hexane) which usually produce biomass and solvent residuals as the by-products. The biomass residues have high protein content and can be used as the feed for animals. It can also be anaerobically digested for production of biogas or be used as the feedstock for biopolymer production [4–6]. Alternatively, recent studies suggest the

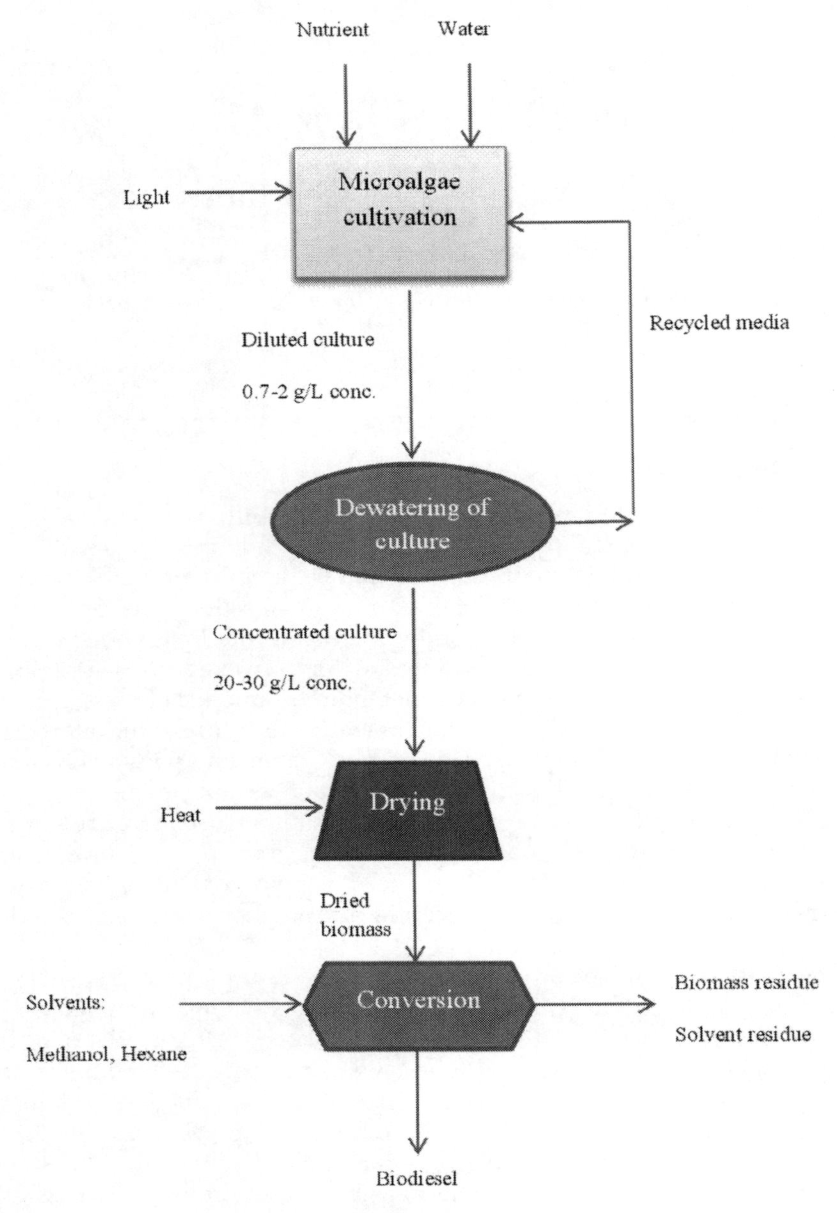

FIGURE 5.1
The flowchart of biodiesel production from microalgae.

direct combustion of biomass residue for production of electricity and high quality fly ash [7–9]. While recent progress in conversion of biomass to valuable products is promising, the high cost of microalgae production still makes it noncompetitive option [10–13].

The different parts of algal biodiesel production (e.g., cultivation of microalgae, harvesting of biomass and conversion of biomass to biodiesel) were described in detail in this chapter.

5.2 Cultivation of Microalgae

5.2.1 Cultivation Systems

There are two types of microalgae cultivation systems used in algal biorefineries: open raceway ponds, and photobioreactors. The schematic of both systems are shown in Figure 5.2. Figure 5.2.a shows the top view of the open raceway pond in which a paddle wheel is used to circulate the culture medium in the reactor. Figure 5.2.b shows the side view of two common types of closed photobioreactors. The culture medium is circulated in the photobioreactors by using a pump (the tubular photobioreactor that is shown on the left) or a sparger (the flat plate photobioreactor that is shown on the right).

In open raceway ponds (Figure 5.2.a) the culture is grown in a shallow pond which is operated in the outdoor condition. In commercial scale, the open pond is constructed directly on the compact land to minimize the cost of construction materials. In this regard, land is scarified, compacted and the berm walls were lined with plastic to control erosion. For pilot scales, the open raceway pond is constructed using concrete or various types of plastic [14]. The cost of construction and life span of the pond can be improved by using the novel polymers with enhanced mechanical properties [15–17]. Open raceway pond is usually facilitated with a paddle wheel to provide sufficient mixing throughout the channel. The mixing prevents the culture from sedimentation, allows uniform light availability and provides sufficient CO_2 mass transfer to the culture. Efficient light supplementation, lowering water need, maintaining sustainable high biomass productivity, and CO_2 supply are operating challenges for cultivation of microalgae in open ponds [18]. The light source for the open raceway pond is usually sunlight, which penetrates into the culture from the top of the reactor. In this regard, even at the locations with high average annual sunlight intensity the culture pigmentation limits the light penetration into depth of the cultures [19]. While the open raceway pond usually works with low depth (10–20 cm), the concentration of microalgae does not usually exceed 1 g/L in open raceway pond. High evaporation loss due to

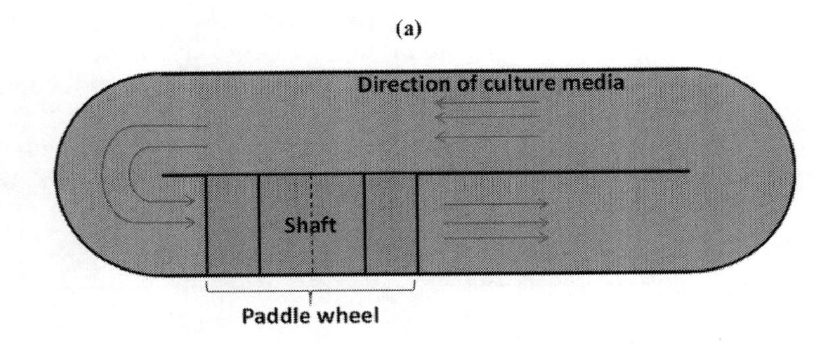

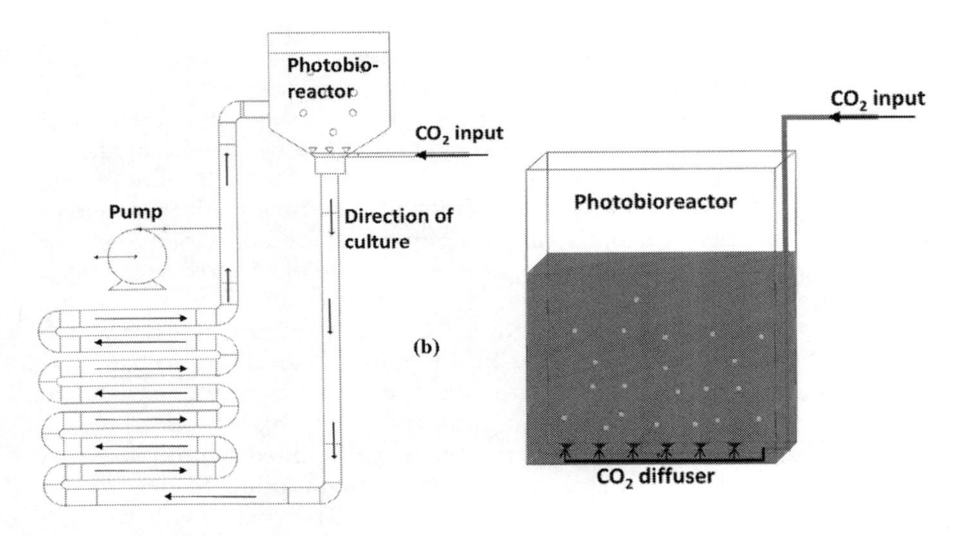

FIGURE 5.2
Schematic of cultivation systems including (a) open raceway pond and (b) photobioreactors.

a large exposure of the culture to the surrounding environment is another drawback for cultivation in open raceway pond. In addition, culture in an open system is usually vulnerable to contamination either from a competing algal species or zooplanktons. The contaminated culture needs to be discarded and replaced with fresh media and inoculum which increases the operating costs of cultivation in open raceway pond. In cultivation of microalgae, the reactors are usually supplied with concentrated CO_2 (an essential macronutrient for micro-algae growth) since the dissolution of atmospheric CO_2 in standard media is low. However, the short traveling time of CO_2 bubbles in open raceway pond lead to inefficient dissolution of CO_2 and high loss of CO_2 to atmosphere in open raceway pond [20]. While the flue gas of power plant can inexpensively provide CO_2, construction of open raceway nearby the industrial plants is challenging as it requires high area of the land where land is usually restricted and expensive.

In photobioreactors, the biomass productivity is enhanced by increasing the light availability the culture. Two common types of photobioreactor were shown in Figure 5.2.b. In a tubular photobioreactor, the culture is pumped into the tubular shape reactor which is exposed to artificial light or sunlight [21]. In a plate photobioreactor, the culture is bubbled from the bottom to prevent the sedimentation and provide sufficient carbon source for microalgae growth. The operation of photobioreactor in a closed condition eliminates the evaporation and CO_2 loss to atmosphere [18]. While the cultivation of microalgae in a controlled photobioreactor allows high biomass productivity with minimum contamination risk and lowers the need for water and CO_2, the high construction and operating costs for photobioreactors are major drawbacks. To commercialize production of biodiesel from microalgae where the cost of construction and operation of cultivation system is critical factor, an open raceway pond regardless of its drawbacks is the only possible system. The application of photobioreactor is limited only to maintenance of inoculum required for cultivation in open raceway ponds as well as pharmaceutical and nutraceutical industries in which the purity of the culture is significantly important [22].

The advantages and disadvantages of two different systems for cultivation of microalgae were summarized in Table 5.1.

TABLE 5.1

Advantages and disadvantages of two conventional cultivation systems (i.e., open raceway pond and photobioreactor) for cultivation of microalgae

	Open raceway pond	Photobioreactor
Biomass productivity	Low	High
Production of desired species	Difficult	Possible
Light penetration	Inefficient	Efficient
Evaporation	Extremely high	Low
CO2 supply	High loss to atmosphere	Minimum losses
Operating cost	• Cost of electricity for the paddle rotation • Low maintenance cost	• Cost of electricity for pumping the culture, light production, oxygen degassing • Low maintenance cost
Construction cost	• Lower construction materials requirement • Low labor cost	• Higher construction materials requirement • High labor cost

5.2.2 Cultivation in Low Quality Water

The possibility of microalgae cultivation in low quality water is one of its advantages relative to terrestrial plants. Some species of microalgae can be cultivated by using brackish water and seawater as the water source [23]. The cultivation using seawater and brackish water can address severe scarcity of fresh water [24–27]. In addition, the saltwater is easily accessible in the locations which have most suitable weather for cultivation of microalgae (e.g., Florida, California) and thereby cultivation in seawater can lower the cost of cultivation. Microalgae species that can produce high amount of lipid in saline water are enlisted in Table 5.2.

In addition to saline water, wastewater is an other sustainable source of water and nutrients for cultivation of microalgae. The essential macronutrients for microalgae growth are C, N and P. In this regard, conventionally the nutrients are supplied in the form of CO_2 and fertilizers. The supplementation of nutrients is critical to environmental and economic sustainability of microalgae [31]. As an alternative method, the wastewater can be used as the source of nutrients for cultivation of microalgae. The cultivation using wastewater also offer a feasible biological wastewater treatment method as it significantly reduces the amount of C, N, and P in the waste. Therefore, it can address the environmental challenges that correspond with discharge of agricultural, municipal and industrial waste streams to water resources.

Some organic compound (e.g. fatty acids and amino acids) can be utilized by microalgae so that the cultivation in wastewater can also induce the mixotrophic growth of microalgae [32]. The mixotrophic growth of microalgae in wastewater can lower the CO_2 supplementation requirement and thereby reduce the cost of microalgae production. High concentration of ammonium in wastewater make the cultivation using wastewater challenging since the conversion of ammonium to ammonia in high pH (>9) can inhibit the growth of microalgae. In this regard, the concentration of ammonia in media is conventionally maintained at low level by controlling the pH of media [33].

TABLE 5.2

Reported microalgae cultivation using salinewater

Cultivation system	Species	Brackish water	Seawater	Productivity	Reference
Open pond	*Nannochloropsis* sp.		×	0.035 g/L/day	[14]
Photobioreactor	*Nannochloropsis salina*		×	25 $g/m^2/day$	[28]
Open pond	*Tetraselmis* sp., *Cyclotella* sp., *Dunaliella* sp.,	×		Not reported	[29]
Open pond	*Nannochloropsis* sp.		×	25 $g/m^2/day$	[30]

Wastewater usually contain high concentration of particulates which increase the turbidity and lower the light penetration to the media [34]. To lower the turbidity of the media, the concentration of particulates is usually decreased through dilution or flocculation before the cultivation. The cultivation is less challenging for mixotrophic cultures since the growth is not very dependent on the light availability. For outdoor cultivation using wastewater, the high chance of contamination by other microorganisms can also negatively affect the productivity of microalgae. The extreme environment such high pH and high ammonia concentration can lower the chance of bacterial contamination [20].

The agricultural wastes including animal manure, dairy waste, olive-oil mill, etc. have high concentrations of nutrients which can be used for cultivation of microalgae [35]. While some industries directly dispose the waste on the land which cause the eutrophication of ground water and lakes, the wastes are sometimes managed through anaerobic and aerobic digestion [7]. The anaerobic digestion of manure reduces microorganisms' population and the concentration of nutrients of the waste effluent, but the N and P concentrations are still sufficient for cultivation of microalgae. In this regard, microalgae cultivation can be considered as a secondary treatment method to further remove organic and inorganic pollutants [35]. The anaerobically digested animal wastewaters usually contain high concentration of ammonium. To reduce the inhibitory effect of the ammonium the media is usually diluted especially for the pH>8. The cultivation in wastewater also makes the production of biomass with a desired composition (e.g. high lipid) difficult since the type of waste can significantly varies the composition of the product [36–38].

The summary of the reported cultivation in different wastewaters is shown in Table 5.3.

Chung et al. (1978) reported 5 $g/m^2/d$ as the productivity and 76% as the N removal for cultivation of *S. platensis* in digested pig waste [39]. They also showed that addition of sodium bicarbonate to the wastewater improved the growth of microalgae. Their results suggested that mixotrophic growth of microalgae in wastewater lead to higher productivity relative to culture grown in standard media. In a more recent study, the cultivation of microalgae in acerbically digested dairy manure was studies by Wang et al. (2010) [40]. Their studies showed the biomass productivity of 0.08 g/L/d, maximum lipid productivity of 11 mg/L/d for cultivation of *Chlorella* sp. Further, they reported 75.7–82.5% from the dairy waste after completion of the cultivation experiments. The cultivation of *Arthrospira* in anaerobically digested cattle manure was studied by Lincoln (1996) [41]. They observed the lower growth of microalgae at higher concentrations of manure suggesting the inhibitory effect of ammonia on the growth. Overall, they reported the maximum biomass productivity of 70 mg/L d and maximum NH_4 removal rate of 24 mg/L d for cultivations in digested cattle manure.

TABLE 5.3

Summary of cultivation of microalgae in various types of wastewater

Microalgae specie	Growth media	Productivity	N removal	Reference
S. platensis	Anaerobically digested pig waste	5 g/m^2/d	76%	[39]
Chlorella sp	Digested dairy manure	0.08 g/L/d	75.7–82.5%	[40]
Arthrospira	Anaerobically digested cattle manure	0.07 mg/L/d	24 mg/L d	[41]
Chlamydomonas reinhardtii	Municipal (centrate)	0.8–2 g/L/d	22–55 mg/L	[42]
S. platensis	Seawater supplemented with effluent from anaerobically digested pig waste	4.8–9 g/m^2/d	100% NH_4-N	[43]
Botryococcus braunii	Municipal (secondary treated)	0.34 g/L/d	Not reported	[32]
S. platensis	Wastewater from the production of sago starch	14.4 g/m^2/d	100% NH_4-N	[44]
Dunaliella tertiolecta	Carpet mill	0.03 g/L/d	99.8% NO_3-N	[45]
Spirulina	Waste from sugar industry	0.32 g/L/d	Not reported	[46]

Kong et al. (2010) have studied the growth of *Chlamydomonas reinhardtii* in undiluted concentrated wastewater [42]. Their studies showed that transition of cultivation system from flask to biocoli resulted in an increase in the biomass productivity from 0.8 g/L/day to 2 g/L/day. Their study also indicates the lipid productivity of 0.5 g/L/day for cultivation in biocoli. Olguín et al. (1997) studied the cultivation of *S. plantesis* in seawater supplied with anaerobically digested pig manure [43]. Their studies indicated the maximum biomass productivity of 9 g/m2/day and 100% ammonium removal from the waste water by using digested pig manure as a nutrient source. Orpez et al. (2009) observed 0.34 g/L/day productivity for cultivation of *Botryococcus braunii* in municipal wastewater (secondary treated) [32]. Their results also indicated ~ 18% wt/wt lipid contents for the biomass cultivated in municipal wastewater.

The cultivation of microalgae in industrial wastewater was also reported in literature. For instance, Phang et al. (2000) was able to obtain 14.4 g/m^2/d productivity by cultivation of *S. platensis* in wastewater from the production of sago starch [44]. Their results also showed that cultivation experiments resulted in ~100% removal of ammonium from the wastewater. Chinnasamy et al. (2010) [45] have used the mixture of 85–90% carpet industry effluents with 10–15% municipal sewage for cultivation of *Dunaliella tertiolecta*. They reported 0.03 g/L/d biomass productivity, ~7% wt/wt lipid contents and 99.8% NO_3-N removal for cultivation using a carpet mill as a nutrient source. Molasses waste from sugar industry (0.25–0.75 g/L) was used for cultivation of *S. platensis* in a study by Adrade and Costa (2007) [46]. They showed the biomass productivity of 0.32 g/L/d can be obtained by using sugar waste for the cultivation of microalgae.

5.2.3 Effect of Nutrient on Lipid/Starch Content

The essential nutrients for microalgae growth are classified as (a) macronutrients; the nutrients which are required in large quantity (e.g., C, N and P) and (b) micronutrients; the nutrients which are supplied in small amounts (e.g., Ca and Mg) [47]. N is the most important nutrient (after C) and is provided in inorganic forms of NO_3 and NH_4 for autotrophic cultivation of microalgae [48]. N is a core element in the structure of proteins; the essential macro-compounds in photosynthesis reactions. The status of N availability in the culture can be expressed as (i) N sufficiency; e.g., N is sufficient for cultivation of microalgae (ii) N limitation; e.g., N is available in the limited amount of microalgae (iii) N deficiency; e.g., N is not available for microalgae growth [49]. While N deficiency inhibits the cell growth, some studies reported that microalgae can still grow when N is limited by relying on the intracellular storage of this macronutrient [2,50].

The change in the status of N in a culture medium can significantly change the composition of the biomass [48]. For instance, while microalgae cells have polar lipids during N sufficiency, the contents of neutral lipid and starch can drastically increase by limiting N in the media [51,52]. Therefore, for production of biomass desired for biodiesel production, the cultivation is conventionally performed in N sufficient media followed by limitation/deficiency of N. Another method for enhancing the lipid productivity is through the dilution of culture media [53]. In addition to N, the limitation and deficiency of other nutrients (e.g., P) can enhance the lipid production from microalgae [54–56]. Also, the other methods for inducing lipids in microalgae can be the increase of salinity and change of the light conditions [57]. The various strategies for producing algal biomass with high lipid are summarized in Table 5.4.

Zhu et.al. (2014) investigated the metabolism changes for biosynthesis of lipid and starch during N starvation of microalgae [58]. They observed that the transition from N sufficient media to N deficient condition resulted in a drastic decrease in the growth of *Chlorella zofingiensis*, a decrease in the chlorophyll content, and an increase in starch and lipid contents. They reported 24.5% as the maximum lipid content of *Chlorella zofingiensis* under the N starved condition. Fernandes et. al. (2013) similarly observed chlorophyll degradation under the nutrient depletion condition coincided with a decrease in the growth and increase in lipid/storage contents of microalgae [59]. Their result also indicated degradation of starch during prolonged cultivation of *Parachlorella kessleri* grown under nutrient depleted media resulted in a further increase in lipid content to a final value of 29%

TABLE 5.4

Summarized strategies reported in literature for enhancement of lipid production from microalgae.

Microalgae species	Strategy for lipid enhancement	Maximum lipid content reported (wt/wt)	Reference
Chlorella zofingiensis	N starvation	24.5%	[58]
Parachlorella kessleri	Nutrient depletion	29%	[59]
***Scenedesmus rubescens*-like**	N/P deficiency	42%	[60]
Not reported	Salinity stress	23.4%	[61]
***Desmodesmus* sp. F2**	Changing environmental/ nutrient conditions	54%	[62]
***M. dybowskii* Y1**	Fluctuating intensity of solar radiation	32%	[63]
***Tetraselmis* sp. M8**	UV-C treatment	65%	[64]

wt/wt. Optimization of the macronutrient (i.e., N and P) supply to obtain the maximum lipid content from *Scenedesmus rubescens*-like microalga was studied by Tan et al. (2014). Their study showed that cultivation in the media low in N (3 mM–10 mM) and P concentrations (20 µM) led to a maximum lipid content of 42%.

In addition to nutrient limitation/deficiency, some studies reported an increase in the lipid productivity by changing the salinity, temperature and light conditions during growth of microalgae. For instance, the effect of salinity stress (0, 0.5, 1, and 2 g/L NaCl) on lipid induction of microalgae was investigated by Venkata Mohan et al. [61]. Their results indicated a maximum lipid content of 23.4% for the culture grown in media with 1 g/L NaCl. The combination of environmental and nutrient stress to obtain maximum lipid in thermo-tolerant microalgae was evaluated by Ho et al. (2014). In their study, they observed 9-day of cultivation under optimum conditions (light intensity of 700 µmol/m^2s, temperature of 35 °C and initial N of 6.6 mM) resulted in a maximum lipid content of 54% in microalga *Desmodesmus* sp. F2 [62]. The effect of intensity fluctuation of solar radiation on the lipid accumulation of *M. dybowskii* Y1 was studied by He et.al. (2015) [63]. Their results showed the highest lipid content of 32% for *M. dybowskii* Y1 and the highest lipid productivity of 35 mg/L/d for *Chlorella* sp. L1 under high fluctuating intensity. Sharma et al. (2014) investigated enhancing lipid production from *Tetraselmis* sp. by applying UV-C radiation [64]. Their results showed that the treatment of microalgae with UV-C significantly enhanced the lipid accumulation and lipid content of UV-treated microalgae reached up to 65%.

The lipid accumulation usually coincides with a decrease in the growth. To address this challenge, cultivation is first performed in N rich media in order to obtain high productivity. The microalgae cells were then exposed to N-stress conditions which cause the conversion of accumulated starch to lipid. In addition to the N status, the cultivation mode can also affect the lipid productivity. A previous study reported that heterothrophic cultivation of microalgae can enhance the production of microalgae. On the contrary, some studies reported that mixotrophic growth of microalgae in wastewater can result in lower lipid contents of microalgae relative to the standard media. The source of N also was reported to affect the lipid productivity. Between various of N sources (NO_3^-, NH_4^+, and urea), nitrate was shown in literature to be the most suitable form of N for production of lipid [65].

Lipid accumulation can occur during N starvation originating from cellular interconversion of carbohydrate to lipid. As reported in previous studies, N deficiency leads to inhibition of cell growth followed by a conversion of storage carbohydrates to triacylglycerol (main source of biodiesel) [52,55]. However, recent studies suggested the possibility of lipid accumulation in the microalgae during growth phase (Figure 5.3). Their analysis showed that

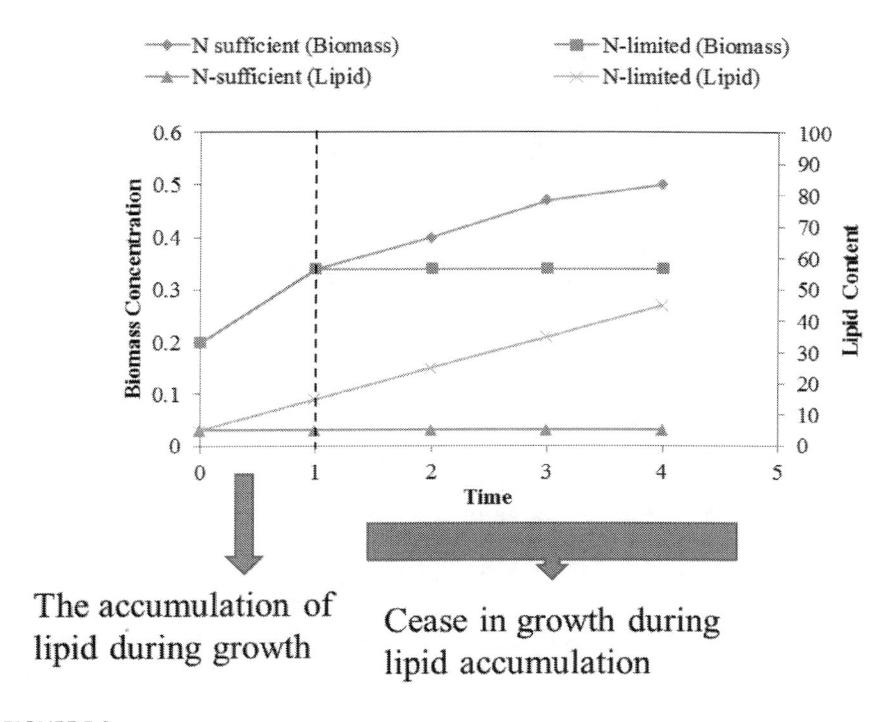

FIGURE 5.3
Hypothetical graph shows literature reported mechanisms for accumulation of lipid in microalgae.

during N limitation in culture media, the cell growth continues for a short period by recycling nitrogen from stored nitrogenous compounds [50,66]. The decrease in the cellular content of N is reported to be coincided with an increase in lipid contents suggesting the *de novo* synthesis of lipid during the growth phase [67].

While recent studies show promising results, the sustainable production of biomass with high lipid contents without compromising the productivity is still a constraint for commercialization of algal biodiesel. To address this challenge, one strategy is to cultivate microalgae with lower nutrient concentrations. However, lowering the nutrient concentration can result in a final low biomass concentration and handling a large volume of culture media for the dewatering process. In this regard, finding an appropriate design which promotes the lipid productivity through optimizing the nutrient and biomass concentrations simultaneously is essential for successful commercialization of algal biodiesel.

5.3 Harvesting of Microalgae

The cost of microalgae production can be reduced significantly by finding a low cost harvesting method, which accounts for 20–30% of the total production costs [21]. There are two methods for harvesting of microalgae: (1) two stage dewatering in which the biomass concentration is increased through primary thickening followed by final harvesting of the concentrated culture and (2) one stage harvesting in which the culture is directly harvested to a final concentration. In one step harvesting, a pre-dewatering process is performed on the culture at low concentration, 0.7–2 g/L [68]. The conventional approaches for initial harvesting are flocculation, coagulation, floatation and sedimentation. An ideal strategy for harvesting of microalgae should allow low cost separation/reuse of the media without any side effect on the quality of biomass [23].

While gravity sedimentation of biomass is low cost method and allow reuse of the media, for majority of conditions the sedimentation is a time consuming process [69]. Due to very small size of microalgae cells and their density close to the water density, the settling rate of majority of microalgae species is in the range of 1–5 cm/hr. The long-time sedimentation of microalgae can result in deterioration of the biomass. To increase the rate of settling, the size of particles is conventionally increased by addition of chemical flocculants/coagulant [68]. The microalgae cells are negatively charged on the surface, which helps their suspension stabilize in the media. One way to increase the settling rate would be through inducing the aggregation of microalgae cells by addition of flocculants. In this regard, the cationic polymers (e.g., cationic polyacrylamides, surfactants and chitosan) bind to the surface of the cells, reduce the charge and induce the formation of large size particles [22]. While the effectiveness of the majority of flocculants decreases in saline water, chitosan was shown to work efficiently even at high salinity. It also was reported to result in low contamination in the harvested biomass [68]. However, the high cost of chitosan limits its application in commercial scales.

In coagulation, a cation binds to the surface of microalgae cells can lead to the neutralization of the surface charge of the cells followed by the destabilization of their suspension [70]. The addition of lime can further assist the sedimentation process through the sweeping flocculation. The typical coagulants are metallic salts such $Al_2(SO_4)_3$ (alum) and $FeCl_3$. The coagulation efficiency can significantly reduce in high pH due to the lowering of formation of hydrolysis products [69]. For an efficient harvesting, a large amount of coagulants is required to be added to the culture which increases the cost of microalgae production [68]. Also, the high amount coagulants will result in contamination of products. Therefore, for food or biofuel application, an additional treatment step is required to improve the quality of biomass. The amount of coagulation added for an efficient harvesting is dependent on the cell structure. Literature reported that the coagulation rate can significantly increase by increasing the content of an extracellular polymeric macromolecule (algogenic organic matter) in the cell wall [71,72].

Centrifugation is another conventional method, besides flocculation/coagulation, for dewatering of microalgae culture. The harvesting using centrifugation is fast, recovers >90% of biomass and allows the reuse of recycled media for further cultivation [73]. However, the process is not cost/energy efficient, which make it feasible only for pharmaceutical/nutraceutical industries. Moreover, the application of high shear forces during centrifugation might cause the cell damage and affect the quality of final products. The microalgae culture can also be dewatered using filtration. Similar to centrifugation, filtration allows high recovery of biomass and recycling of the media for further cultivation. However, the accumulation of biomass on a membrane during filtration process can increase the pressure drop and decrease the filtration rate [68]. In this regard, the backwash step is applied after reaching the lowest standard flow rate to prevent the damage to the membrane. The filtration can be a feasible method for harvesting of the species with larger cell size as the application of small-size membrane filters has higher maintenance/operating costs [69]. While filtration can be a suitable harvesting method for small scale operations and for the cells sensitive to shear stress, both centrifugation and filtration are usually used as a second step for the two-step harvesting strategy.

On the contrary to flocculation/coagulation, floatation increases the lifting force to allow the particles to rise to the surface of the media [22]. This method is typically used in wastewater treatment for removal of suspended particles [74]. Prior to floatation, addition of coagulants is essential for the efficient attachment of air bubbles to the particles. The floatation was reported to be a promising strategy for harvesting of microalgae. However, the efficiency of the floatation can significantly decrease by an increase in salinity of media suggesting the infeasibility of this method for harvesting of marine cultures [68]. The most common flotation technique is dissolved air floatation (DAF) in which the cells are harvested from the surface of culture [74]. In dispersed ozone flotation (DOF), bubbles are charged on the surface allowing more intense binding force between microalgae cells and bubbles and faster floatation [69]. A similar strategy was performed for electrical harvesting approaches [68]. In this regard, thanks to the cells' negatively charged surface, they were accumulated on the surface or the bottom of the culture by applying electrical field. In the electrical harvesting approaches, microalgae cells are attached to a positively charged anode resulting in neutralization of their surface charge and formation of cell clumps. While both aluminum- and iron-made anodes can be used in this process, aluminum showed better efficiency due to the higher current efficiency [74]. Continuous harvesting using the electrical techniques requires regular maintenance as fouling can reduce the process efficiency [75]. The electrical harvesting efficiency is strongly dependent on the current density and mixing rate of the culture. Also, an increase in cell density and pH can reduce the efficiency of process. On the contrary to other harvesting techniques, the electricity-based harvesting seems to be more efficient for the cells grown in seawater. For better dewatering results, the electrical harvesting is typically combined with floatation. Overall, the electricity-based harvesting requires expensive equipment, is highly energy consuming, and is not suitable for application to a large-scale production of biodiesel.

5.3.1 Bioflocculation and Autoflocculation

Bioflocculation is the harvesting process in which addition of secondary microorganisms or their excreted chemicals cause flocculation of the microalgae cells. This method is more economical and environmentally sustainable and it allows reuse of the recovered media. Moreover, no chemicals are added during bioflocculation so that additional pretreatment processes before biodiesel production are not necessary. The addition of plant derived biopolymers (e.g., polysaccharide derivatives) is one of the strategies for bioflocculation. The process is similar to the flocculation using synesthetic chemicals, but it involves addition of low cost and degradable substances to the culture. The interaction of organic substances (e.g., pectin, mannose, cellulose) with microalgae cells results in neutralization of the cell surface charge and formation of cell clumps in the culture. The process is then followed by sedimentation in which the biomass cake is removed from the bottom of settling tank.

The microbial bioflocculation is another strategy that is associated with addition of microorganisms to the microalgae. Some microorganisms are able to produce extracellular polymeric substances (EPS) which act as a natural flocculent. This process is less expensive and more environmentally sustainable relative to addition of synthetic flocculants. However, the efficiency of the process and the amount of the flocculent is highly species-specific. Moreover, the co-culturing of microalgae with another microorganism can affect the productivity and quality of the biomass and the by-products. The genetic modification of microorganisms is a promising strategy to address some of these challenges. Fungus-microalgae co-cultivation is another promising strategy that results in simultaneous formation of microalgae cells pellets along with fungal pellets (due to their surface charge) and efficient harvesting of the culture. Microalgae can be co-cultivated with fungus through mutual symbiotic communication and thereby their co-cultivation does not affect the productivity of microalgae. Moreover, Zhang et al. [76] suggested that co-cultivation of microalgae with oleaginous filamentous fungus not only reduce the harvesting cost, but also contribute to overall lipid productivity of final product.

Table 5.5 summarizes the results of previous studies reported on the bioflocculation of microalgae. Hamid et al. (2014) studied the harvesting of microalgae *chlorella* sp. using *Moringa oleifera* (MO) seed and its derived protein. Their analysis showed that addition of 10 mg/L MO seeds resulted in a 97% removal rate which is partially higher than the removal rate by addition of 10 mg/L MO protein (95%). Their study suggests that application MO seeds instead of a conventional flocculent (i.e., aluminum sulfate) lead to higher removal efficiencies, biomass recovery and thus it is a promising alternative technique for economic and environmental sustainable harvesting of microalgae. The harvesting of microalga *Chlorella vulgaris* UMN235 through co-cultivation with fungal str. *Aspergillus oryzae* was studied by Zhou et al. (2013) [78]. They observed 100% harvesting efficiency of heterotrophic cultures by using 1.2×10^4 spores/mL of fungi and optimum conditions (20 g/L glucose, pH of 4.0–5.0). Futhermore, for autotrophic cultures (without addition of glucose), their results showed a partially lower harvesting ratio (93%). In another study by Al-Hothaly, bioflocculation of microalgae *Botryococcus braunii* with *Aspergillus fumigatus* was investigated. Their results suggest that optimized harvesting achieved

TABLE 5.5

Summary of results for bioflocculation of microalgae reported in literature.

Microalgae species	Flocculation method	Removal Efficiency	Reference
***Chlorella* sp.**	*Moringa oleifera* seed powder	97%	[77]
***Chlorella* sp.**	*Moringa oleifera* protein powder	95%	[77]
***Chlorella vulgaris* UMN235**	Fungus pelletization	100%	[78]
Botryococcus braunii	Fungus pelletization	98%	[79]
***Chlamydomonas* sp.**	Cationic cassia	84%	[80]
***Chlorella* sp.**	Cationic cassia	92%	[80]

through the 1:40 ratio (microalgae: fungi) over 12 h of experiment which resulted in 98% harvesting of microalgae. Banerjee et. al. (2014) studied the harvesting of microalgae using a naturally derived biopolymer (cationic cassia) [80]. Their results show the flocculation efficiency of 84% for *Chlamydomonas* sp. CRP7 using 80 mg/L biopolymer. Also, they observed a higher harvesting efficiency (92%) for *Chlorella* sp. CB4 when the culture was supplied with less amount of biopolymer (35 mg/L).

Autoflocculation is another low cost harvesting strategy, besides bioflocculation, in which the microalgae cells spontaneously form flocs when the pH of media is suddenly increased typically pH value 10, and when $Ca(OH)_2$ and $Mg(OH)_2$ precipitates are triggered to form. These precipitates are postulated in literature to bond with anionic algal carboxylate groups and cause "autoflocculation" of microalgae cells [81,82]. Also, for specific microalgae species, the microalgae cells can excrete extracellular polymeric substances (EPS), which cause bridging between the cells and triggering formation of flocs through autoflocculation. Therefore, addition of species with the capability of producing bioflocculent can be an alternative to bioflocculation through co-culturing with another microorganism (fungi or bacteria). On the contrary to flocculation, autoflocculation is a less expensive harvesting strategy and the recovered media can be reused for cultivation without any pretreatment as no synthetic flocculent is required for this method. Autoflocculation is conventionally occurred by increasing the pH of the media through addition of NaOH. Increasing the pH is specifically advantageous for the culture grown using wastewater as it can reduce the population of competing microorganisms. The summarized results of previous studies on microalgae autoflocculation is reported in Table 5.6.

TABLE 5.6

Summary of previous studies on microalgae autoflocculation.

Method	Species	Efficiency	Reference
Increasing pH	Freshwater microalgae (*Chlorella vulgaris, Scenedesmus* sp., *Chlorococcum* sp.) and marine microalgae (*Nannochloropsis oculata, Phaeodactylum tricornutum*)	90%	[83]
Increasing pH	*Phaeodactylum tricornutum* and *Scenedesmus* cf. *obliquus*	~70–85%	[84]
EPS production	*Chlorella vulgaris, Neochloris oleoabundans*	31–50%	[85]
EPS production	*Ettlia texensis*	90%	[86]
EPS production	*Chlorella Vulgaris*	71%	[87]

In the study by Wu (2012), flocculation of marine microalgae at high pH>10.5 was investigated [83]. They reported the harvesting efficiency of 90% for a variety of microalgae species (see Table 5.6) by increasing the pH of media from 7.5 to 11.5. The mechanism was suggested to be a hydrolysis of Mg^{2+} in the media to form the magnesium hydroxide precipitate, which leads to surface charge neutralization on the microalgae cells, and thereby, to coagulation of microalgae cells. Also, their study indicated that polysaccharide production from microalgae could also be another reason for the efficient autoflocculation of microalgae by increasing the pH. The reuse of media was proven to be possible through neutralization of the pH and supplementation of essential nutrients. Spilling et al. studied the effect of pH, turbulence and cell density on autoflocculation of two species of microalgae (*Phaeodactylum tricornutum* and *Scenedesmus* cf. *obliquus*) [84]. Their results indicated that pH is the most important factor affecting harvesting of microalgae. The threshold of pH for flocculation is shown to be species-specific and reported as 11.3 for *Scenedesmus* cf. *obliquus* and 10.5 for *Phaeodactylum tricornutum*. Also, they reported higher autoflocculation efficiencies for *Phaeodactylum tricornutum* (~85%) relative to those for *Scenedesmus* cf. *obliquus* (~70%). They suggested that one possible strategy for increasing the pH and harvesting of microalgae at the same time would be through manipulating the CO_2 feeding cycles during the growth cycle.

Autoflocculation during the cultivation of microalgae through excretion of EPS from microalgae cells were also evaluated consistently in the literature. For instance, Salim et al. (2012) studied the increasing harvesting rate of *Chlorella vulgaris, Neochloris oleoabundans* through co-cultivation with microalgae species which have the potential for production of EPS (i.e., *Ettlia texensis, Ankistrodesmus falcatus, Tetraselmis suecica* and *Scenedesmus obliquus*) [85]. Their results indicated the maximum harvesting efficiency of 40% for *Chlorella vulgaris* by addition of Ettlia *texensis* and 50% for *Neochloris oleoabundans* through cultivation with *Tetraselmis suecica*. Later, they additionally evaluated the autoflocculation of *Ettlia texensis*was [86] and reported 90% recovery during 3h of sedimentation. Their study demonstrated the production of EPS as a main chemical affecting cell-cell adhesion, and thereby, achieved higher sedimentation rates. The content and composition of fatty acid for the final biomass was also reported in their study to be desirable for biodiesel production. Such observations by Salim et. al. (2013) was later substantiated by Shang et. al. (2015) through the simulation of EPS excretion from microalgae with an addition of glycine. Their study showed that an increase of EPS in the media reduced the autoflocculation at high light intensities (250 $\mu mol/m^2 \cdot s$), but it increased the harvesting efficiency at low light intensities (125 $\mu mol/m^2$ s). They also observed that increasing the mixing rate resulted in more efficient flocculation. Overall, they reported a 70% recovery as the maximum harvesting efficiency, which was obtained from a 3-day experiment for the culture supplied with 0.1 g/L glycine and under 250 $\mu mol/m^2 \cdot s$.

5.4 Drying

In biodiesel production processes, harvesting of microalgae is typically followed by drying of collected slurry. Drying of microalgae is challenging and is a major bottleneck in production of low-cost biomass. In this regard, majority of studies suggest the extraction of product from wet biomass to reduce the overall cost of production. However, in current commercial production of microalgae, drying is necessary prior to conversion of biomass to products since it can significantly affect the yield of conversion [88]. An ideal method for drying microalgae should be an energy efficient process that allows sustainable drying of microalgae without affecting the composition of the biomass. Drying microalgae can be performed using rotary drying, sun drying, spray drying, flashing drying, etc. Selection of drying methods can vary based on the scale of process and the application of products. For biodiesel production, finding an economical sustainable choice for drying microalgae can significantly affect the cost efficiency of production process.

Solar drying is the cheapest and the most feasible drying strategy especially in areas with high hours of sunlight (which typically are the most suitable locations for cultivation of microalgae). For valorization of solar energy as the heat source for drying, a thin layer of biomass is exposed to sunlight. Overexposure of biomass to sunlight may lead to degradation of chlorophyll and to undesirable changes in biomass quality (e.g., color) [89]. So, in order to maintain the quality of biomass, it is necessary to optimize the duration of biomass exposure. Application of sun drying to microalgae is limited to the use of algae for animal feeding and biodiesel production because the long-time drying may deteriorate the biomass resulting in undesirable quality of final products for human use [89]. Alternatively, the solar energy can be utilized indirectly through glass plate solar collectors which are usually designed tubular or flat. The application of a solar dryer can control the heat intensity during the process, which results in better quality products than direct sun drying. However, the composition of product can still be altered during a long-time drying process. Another low-cost method for drying microalgae biomass is cross-flow drying, in which drying is performed by blowing hot air to a layer of biomass. The cross-flow drying is faster than sun drying but may lead to the change in biomass composition without proper time management. The cross-flow drying is not dependent on the weather and is more viable than sun drying for the locations where sun radiation is not enough [74].

Spray drying and rotary drying are more expensive methods than sun drying and cross-flow drying [74]. For rotary drying, biomass is dried by moving in a high temperature tilted cylinder. The use of rotary dryer leads to sterilization of biomass and weakening of the cell membrane of microalgae allowing the efficient extraction of chemicals from dried biomass

[89]. For rotary drying, electricity is used for both heat generation and rotation of dryer. The energy efficiency of the process is highly dependent on the water content of input biomass. In this regard, to obtain a cost-efficient drying process, the water content of input biomass is required to maintain at low level. For rotary drying, application of steam for drying may result in higher energy efficiency relative to electricity. In spray drying, the input biomass is converted to small droplets using an atomizer. The produced droplets are then exposed to hot air in a vertical tower and rapidly dried. The spray dryer is the most expensive method and the quality of biomass is suitable for human consumption. Therefore, the spray drying is typically used for the production of biomass and other products used in pharmaceutical and nutraceutical industries [90].To reduce the cost of drying, the wasted heat from flue gasses of industries can be recovered for drying of microalgae. In this regard, the heat exchangers are used in the location of the points in which the produced heat is the waste heat [91]. The integration methods (e.g. pinch analysis) are typically used to design the optimized heat utilization with maximum energy efficiency.

Specifically for biodiesel production, Guldhe et. al. (2014) investigated the effect of drying methods on the lipid yield of *Scenedesmus* sp [92]. In this regard, biomass was dried using oven-drying, sun drying and freeze drying. The yield was evaluated using microwave assisted solvent extraction and sonication assisted solvent extraction. The longest drying time was for sun drying (72 hr) followed by freeze drying (24 hr) and oven drying (12 hr). Their results showed insignificant effect of drying methods on the yield of lipid suggesting that sun drying can be applied for biodiesel production as it is most simple and cost-effective drying strategy.

5.5 Conversion of Biomass to Biodiesel

In a microalgae-based biorefinery, chemical and thermochemical methods are typically used for conversion of biomass to biodiesel. For chemical conversion method, lipid is extracted by addition of an alcoholic solvent and is converted to biodiesel through transesterification reaction [93]. Transesterification reaction process is shown in Figure 5.4. In this reaction, the alcohol is reacted with triglyceride (lipid) resulting in replacement of glycerol groups with alkyl groups from alcohol. The product of this reaction is fatty acid methyl ester (FAME) which is the final product and the glycerol which is the by-product. The transesterification reaction is typically catalyzed using an alkaline or acidic catalysis [94]. The alkaline catalysis is not desirable if the lipid has high content of free fatty acid as it can lead to saponification of the product. However, for microalgae oil as the lipid has low content of free fatty acid the application of both acidic and alkaline catalysis is possible.

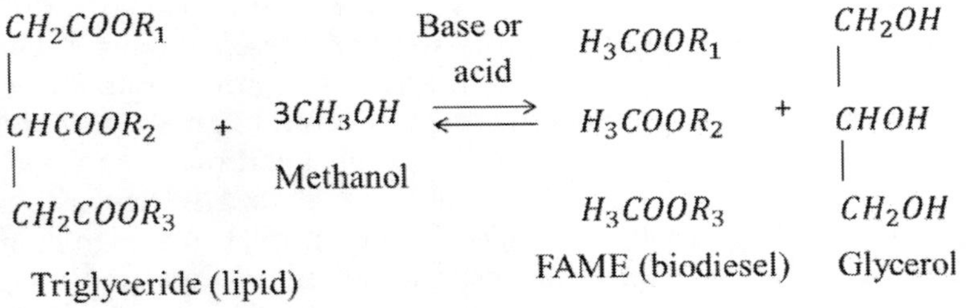

$$\begin{array}{lcccccc} CH_2COOR_1 & & & \text{Base or} & H_3COOR_1 & & CH_2OH \\ | & & & \text{acid} & & & | \\ CHCOOR_2 & + & 3CH_3OH & \rightleftharpoons & H_3COOR_2 & + & CHOH \\ | & & \text{Methanol} & & & & | \\ CH_2COOR_3 & & & & H_3COOR_3 & & CH_2OH \\ \text{Triglyceride (lipid)} & & & & \text{FAME (biodiesel)} & & \text{Glycerol} \end{array}$$

FIGURE 5.4
Transesterification of lipid to biodiesel in which R is hydrocarbon groups.

The chemical conversion requires the purification of final product, the separation from solvents, handling the solvents and residuals. The residual biomass of the process can be converted to biofuel through anaerobic digestion. The output solvents may be purified and reused for the conversion. An alternative to two step chemical conversion (extraction and esterification), *in situ* transesterification can simplify the process by combining the steps [95]. In this regard, mixture of biomass and acidified/alkalified methanol produce FAME during 1hr reaction under temperature of 80–90°C. As shown in Figure 5.5, the *in situ* transesterification process results in two immiscible liquid layers in which the bottom layer contains methanol, catalyst and biomass residual and the top layer has the mixture of hexane and FAME. The *in situ* transesterification can also be performed for the wet biomass if the excessive amount of methanol was used for the process. However, the cost of the process is still high as the output

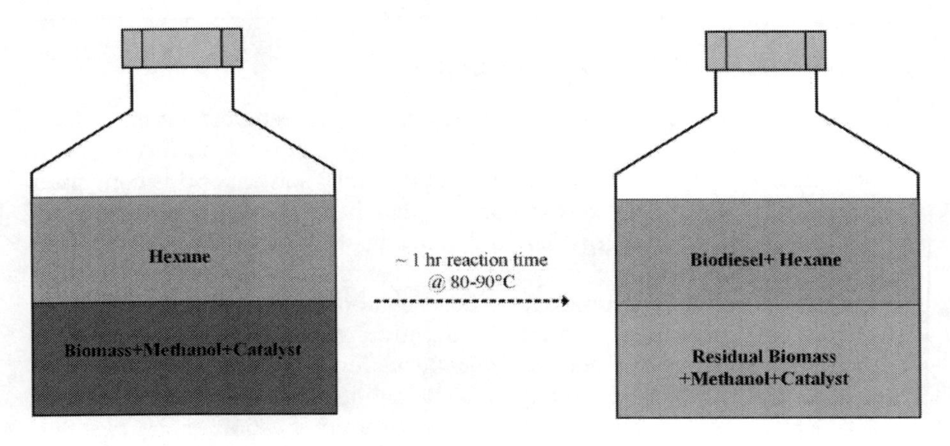

FIGURE 5.5
Overview of *in situ* transesterification of microalgae.

methanol has water and biomass residual which need to be purified prior to recycling. The separation of the residual biomass can be performed using centrifugation or filtration. The water separation from methanol can be performed in a distillation column. As mentioned earlier, the residual biomass can be anaerobically digested for production of biogas which can supply the part of the energy for post treatment of the solvent. The use of end-product of digestion as the fertilizer for cultivation of microalgae can reduce the overall cost of production.

Thermochemical conversion methods reduces/eliminates the application of solvents and therefore is likely more environmentally sustainable. Some thermochemical conversion methods (e.g. hydrothermal liquefaction) have the potential to convert the entire biomass to biodiesel [96]. Hydrothermal liquefaction (HTL) process is the conversion of biomass to biodiesel under high temperature (>300°) and pressure (>2000 psi) [97]. The HTL process is performed in the presence of water. Therefore, the wet biomass from harvesting can be directly used which reduces the cost of microalgae production. The HTL process is a combination of pyrolytic mechanisms at high temperature and hydrolysis in the presence of water. The products of HTL are in the form of solid, aqueous, and gas [98]. The gaseous products are separated from the product mixture by de-pressurizing. The aqueous phase of product is mixture of various products including water, biocrude, and bio-oil. The primary separation of products in an aqueous phase is through gravity separation as the products are immiscible with each other. The posttreatment process is essential for separation of bio-oil and biocrude from water. Biochar is a by-product of HTL and has the application as the flocculants to harvesting of microalgae [99]. Alternatively, it can be sold as a byproduct to wastewater/water treatment industries which can improve the economic efficiency of biodiesel production from microalgae.

Biocrude has high viscosity and is a mixture of cyclic ketons, variety of acids/alcohols, and phenols. The produced biocrude from HTL has high contents of oxygen, nitrogen and sulfur, and thereby, does not have the suitable quality to be used as a fuel source [98]. To improve the quality (e.g., flow properties) of HTL products, the process can be catalyzed using alkaline (e.g., KOH or Na_2CO_3) or acidic catalysts (e.g. acetic acid) [100]. Hydrogenation of the products can also be used for improvement of the product quality by conversion of S, N and O to H_2S, NH_3 and H_2O. The water layer of HTL products contains high content of C and N and can be recycled for cultivation of microalgae. However, Edmundson et al. (2017) observed the inhibitory effect of nitrogenous compounds in the HTL product on the growth of microalgae [97].

The advantages and disadvantages of chemical conversion and HTL are summarized in Table 5.7. The main advantage of chemical conversion is the characteristics of the product which has same quality as diesel. However, as discussed earlier, the product of HTL has high contents of

TABLE 5.7

Advantages and disadvantages of chemical and HTL conversion methods for production of biodiesel

Conversion methods	Advantages	Disadvantages
Transesterification	• The product has the quality of biodiesel • The reaction temperature and pressure are significantly lower than HTL • The product residuals can be reused after relatively easier treatment process	• It is not efficient if the biomass has low content of lipid • The solvents need to handled • It is sensitive to water content of biomass
HTL	• Higher productivity than transesterification as the whole biomass is converted to biocrude • The process eliminates the biomass drying step in microalgae production • The process is faster relative to chemical conversion	• The additional treatment on produced biocrude is necessary prior to use for transportation • It require design and operation of the reaction at very high temperature and pressure which needs the application of an expensive reactor, high operating cost and safety consideration • The nutrient recycling is limited as some of the compounds in the product has negative effect on the growth of microalgae

O, N and S, which need additional processing prior to be used as a fuel. The chemical conversion needs a high amount of solvent, which needs to be handled with caution. While the thermochemical conversion eliminates the use of solvents, the operation of HTL reactors at intensive pressure/temperate is challenging due to high cost of construction and operation. Overall, HTL processing is a promising technique as it allows the conversion of whole wet biomass after harvesting but has not been developed enough yet for commercial scale operation due to operational limitations.

5.6 Microalgal Biodiesel Properties

The fatty acid profile of three different species of microalgae as well as palm, soybean, sunflower, and tallow oil are compared in Table 5.8.

TABLE 5.8

Fatty acid profiles of biodiesel produced from 3 species of microalgae (*S. obliquus* SJTU-3, *Chlorella* sp. TISTR 8990 and *Nannochloropsis oculata*), palm, soybean, sunflower and tallow

Fatty acid	*S. obliquus* SJTU-3 [101]	*Chlorella* sp. TISTR 8990 [102]	Nannochloropsis oculata [103]	Palm [104]	Soybean [104]	Sunflower [104]	Tallow [104]
12:0	0.28					0.5	
14:0			6.8	1	0.1	0.2	
16:0	22.33	22.3	22.7	42.8	11	4.8	23.3
16:1	0.25	1.6	21.5		0.1	0.8	0.1
16:2	0.92						
16:3	6.07						
18:0	0.93	4.5	0.6	4.5	4	5.7	19.3
18:1	1.15	24.2	13.4	40.5	23.4	20.6	42.4
18:2	13.43	44.4	4.9	10.5	53.2	66.2	2.9
18:3	48.44	2.9		0.2	7.8	0.8	0.9
20:0	0.16				0.3	0.4	
20:1	0.47						
20:4			4.9				
20:5	5.36		24.2				
22:0					0.1		

The properties of biodiesel produced from microalgae are species-specific, but the produced FAMEs typically have long hydrocarbon chains with high degree of unsaturation. The biodiesel with high degrees of unsaturation tends to oxidize faster, and therefore has low oxidation stability [105]. However, the large content of unsaturated FAMEs is desirable for flow properties and result in lower cloud points and cold filter plugging points [106]. Further, the cetane number of fuel increases by increasing the degree of unsaturation [107]. High cetane number is the indication of more efficient and faster combustion of biodiesel in a combustion chamber [105]. Density and viscosity of biodiesel are dependent on the degree of saturation and size of carbon chain. Generally, the lower degree of saturation and presence of longer carbon chain fatty acids leads to higher density of biodiesel [105]. Density and cold flow properties of biodiesel from microalgae are comparable to jet fuel and makes it a suitable alternative diesel to be used in aviation industry [108].

The standard quality of biodiesel is reported by the US Department of Energy for B100 (100% biodiesel) and B6 to B20 (6% to 20% biodiesel). The comparison between the standard specification of biodiesel fuel with properties of microalgae biodiesel and petroleum diesel is shown in Table 5.9. The reported data in the literature shows that microalgae biodiesel is

TABLE 5.9

Biodiesel and diesel properties as well as specifications for B100 and B6 to B20 provided by the US Department of Energy[a]

Property	Microalgae biodiesel	Diesel	B100	B6 to B20
Flash point, °C, min[109]	115	60–80	93	52
Cetane number, min[95, 109]	60.73	51	47	40
Oxidation stability, hrs, min[110]	8.83	17.3	3	6
Density at 15 °C, kg/m^3[110]	919	836		
Pour point (°C) [109]	−12	−35 to −15		
Viscosity [109]	5.2	1.2–3.5		1.9–4.1

[a]ASTM D6751-15ce1, Standard Specification for Biodiesel Fuel Blend Stock (B100) for Middle Distillate Fuels, ASTM International, West Conshohocken, PA, 2015, www.astm.org.

a promising alternative fuel source, but possibly needs additional modification to meet the standard specifications.

The cultivation conditions of microalgae can alter the produced biodiesel properties. For instance, Mandotra et al. showed that increasing the concentration of phosphorus in the media initially decreased the cetane number of biodiesel (until 40 mg/L) and increased the degree of unsaturation [111]. However, further addition of phosphorus resulted in an increase in the cetane number and a decrease in the degree of unsaturation. Their data also showed fluctuation of the cetane number, and the degree of unsaturation by pH variation in the media, which suggested a maximum degree of unsaturation of 110.04 and a minimum cetane number of 43.52 for the culture grown at pH = 7. Their results for increasing the light intensity on the culture (from 3000 to 6000 Lux) showed a continuous decrease in the degree of unsaturation (from ~82 to ~26) and an increase in the cetane number (from 50 to 66).

In another study by Ruangsomboon (2015) [112], the properties of biodiesel from microalgae grown in various media, nitrogen source/concentration and cultivation strategies were evaluated. Their results indicated that cultivation in various standard media did not affect the cetane number of biodiesel but altered the cold filter plugging point of the fuel. Also, their data suggested that between various sources of N (i.e., KNO_3, NaNO3, $Co(NH_2)_2$, NH_4HCO_3), application of sodium nitrate as a nutrient source resulted in a maximum cetane number (52.49) and potassium nitrate resulted in lowering of the cold filter plugging point (-1.56°C). Between various nitrogen concentrations, the cultivation with 0.63 g/L N resulted in an optimum cetane number and cold filter plugging point of the fuel (39.74 and 1.44°C, respectively). Additionally, the two-step cultivation strategy did not significantly change the cold flow properties of fuel, but slightly altered the cetane number of biodiesel.

Application of microalgal biodiesel instead of diesel affects the performance and the emission of combustion engine. The reported data by Tüccar (2014) shows that addition of microalgae biodiesel to diesel increases NOx production possibly due to the higher content of oxygen in biodiesel than regular diesel. However, the lower CO emission was observed since the higher oxygen content of biodiesel resulted in more complete combustion of fuel [113]. Their results showed insignificant difference between smoke opacities of biodiesel and diesel.

In another study by Makarevičienė (2014) the performance of diesel fuel mixed with microalgae biodiesel were evaluated and the results were compared with an ordinary biodiesel (rapeseed biodiesel) [114]. While their results indicate ~6% lower calorific values than diesel, which resulted in more fuel consumption, the mixture of biodiesel and diesel showed 2.5–3% higher efficiencies of energy utilization than diesel. Furthermore, their results showed that addition of microalgae biodiesel to regular diesel made it possible to lower the smokiness and hydrocarbon (HC) emissions by 10–75% and by 5–25%, respectively, relative to diesel. Their data suggested that the diesel-microalgae biodiesel mixture produced ~10% lower HC emissions than diesel-rapeseed diesel mixture. However, the application of microalgae biodiesel resulted in ~10% higher smokiness relative to rapeseed biodiesel.

5.7 Conclusion

Production of biodiesel from microalgae is promising because microalgae has the capability of accumulating a high amount of lipid. However, biodiesel from microalgae is not competitive with diesel, and other conventional biodiesels in the market since its production requires large costs for cultivation, harvesting and conversion. Figure 5.6 illustrates a diagram of the suggested path for an economically/environmentally sustainable production of biodiesel. Between various cultivation strategies, cultivation in an open pond using sunlight seems to be the most cost-efficient technique for production of biomass. For cultivation, use of low-quality water (e.g., wastewater, saltwater) as a water source and industrial/municipal effluents as nutrient sources can significantly enhance the economic and environmental sustainability of process. Recycling the water and nutrients from harvesting and conversion processes is essential to reduce the overall cost of production.

Dewatering of microalgae is challenging as the culture concentration is very low after cultivation (0.7–1 g/L), which requires handling of large volume of culture. The conventional strategies (e.g., centrifugation and

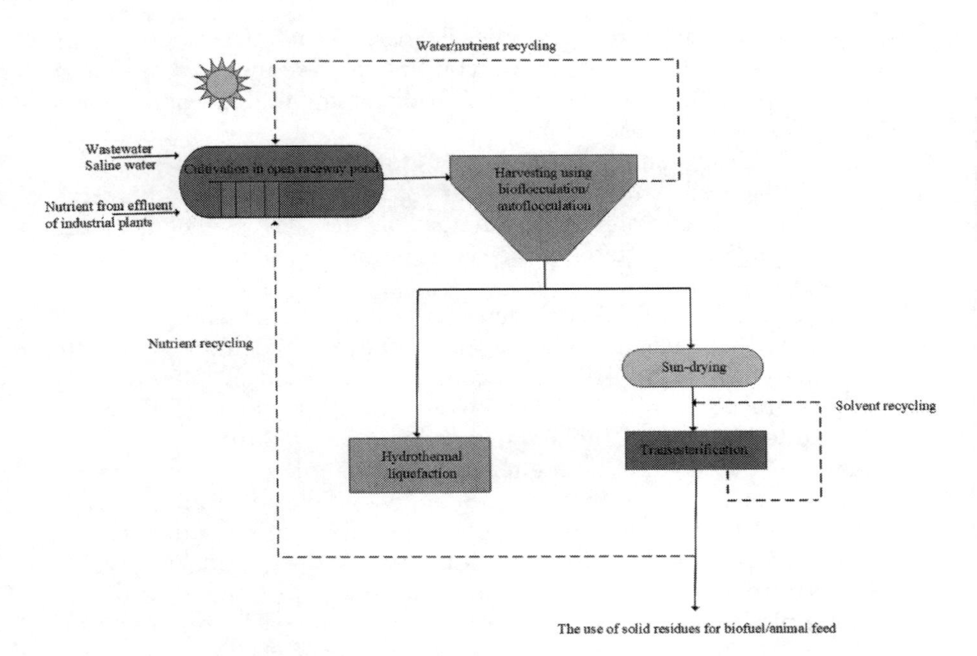

FIGURE 5.6
Schematic of the process suggested for production of biodiesel from microalgae.

filtration) are not energy efficient and not feasible for commercial harvesting of microalgae. While flocculation using chemicals is costly and does not allow recycling of the media, bioflocculation and auto-flocculation are regarded as a promising technique as it does not involve additional synthetic chemicals to the media, and thereby, is a low-cost method allowing the reuse of the media. Between the various drying techniques, sun-drying is time consuming and needs a large area of the land. However, sun-drying is the most suitable method especially for the plants located in the desert (where land is not a limited) due to its significantly lower costs relative to other strategies.

Conversion of microalgae through transesterification results in production of fuel with comparable characteristics to diesel that can be directly used for transportation. However, the chemical conversion methods require application of a large volume of solvents and are not an economically/environmentally efficient method especially when the lipid content of biomass is low. HTL is an alternative strategy for production of biodiesel allowing the conversion of the whole biomass and elimination of the drying step in biodiesel production. However, HTL process requires a reactor that operates at high pressure and temperature. It has limitations in recycling of the nutrients from effluent of conversion.

References

1. M. Hanifzadeh, O. Tavakoli, *Carbon dioxide mitigation using microalgae,* The Inaugural Pacific Rim Energy & Sustainability Congress, 2012.
2. B. Pendyala, A. Vadlamani, S. Viamajala, S. Varanasi, M. Hanifzadeh, *High Yield Algal Biomass Production Without Concentrated CO_2 Supply Under Open Pond Conditions,* Google Patents, 2017.
3. M.M. Hanifzadeh, M.H. Sarrafzadeh, O. Tavakoli, *Carbon dioxide biofixation and biomass production from flue gas of power plant using microalgae,* 2012 Second Iranian Conference on Renewable Energy and Distributed Generation, IEEE, 2012, pp. 61–64.
4. S.A. Rodriguez, E. Weese, J. Nakamatsu, F. Torres, *Development of biopolymer nanocomposites based on polysaccharides obtained from red algae chondracanthus chamissoi reinforced with chitin whiskers and montmorillonite,* Polymer-Plastics Technology and Engineering, 55 (2016) 1557–1564.
5. K. Majdzadeh-Ardakani, S. Zekriardehani, M.R. Coleman, S.A. Jabarin, *A Novel Approach to Improve the Barrier Properties of PET/Clay Nanocomposites,* International Journal of Polymer Science, 2017 (2017) 10.
6. S. Zekriardehani, S.A. Jabarin, D.R. Gidley, M.R. Coleman, *Effect of chain dynamics, crystallinity, and free volume on the barrier properties of poly(ethylene terephthalate) biaxially oriented films,* Macromolecules, 50 (2017) 2845–2855.
7. M. Hanifzadeh, Z. Nabati, P. Longka, P. Malakul, D. Apul, D.-S. Kim, *Life cycle assessment of superheated steam drying technology as a novel cow manure management method,* Journal of Environmental Management, 199 (2017) 83–90.
8. P. Khoshnoud, S. Gunashekar, M.M. Jamel, N. Abu-Zahra, *Comparative analysis of rigid PVC foam reinforced with class c and class f fly ash,* Journal of Minerals and Materials Characterization and Engineering, 2 (2014) 554.
9. N.H. Abu-Zahra, P. Khoshnoud, M. Jamel, S. Gunashekar, *Enhanced thermal properties of rigid PVC foams using fly ash, international science index,* Materials and Metallurgical Engineering, 9 (2015).
10. R. Slade, A. Bauen, *Micro-algae cultivation for biofuels: cost, energy balance, environmental impacts and future prospects,* Biomass and Bioenergy, 53 (2013) 29–38.
11. S. Golbad, P. Khoshnoud, N. Abu-Zahra, *Synthesis of 4A zeolite and characterization of calcium-and silver-exchanged forms,* Journal of Minerals and Materials Characterization and Engineering, 5 (2017) 237.
12. P. Khoshnoud, J.C. Wolgamott, N. Abu-Zahra, *Evaluating recyclability of fly ash reinforced polyvinyl chloride foams,* Journal of Vinyl and Additive Technology, (2016).
13. M.M. Jamel, P. Khoshnoud, S. Gunashekar, N. Abu-Zahra, *Mechanical properties and dimensional stability of rigid PVC foam composites filled with high aspect ratio phlogopite mica,* Journal of Minerals and Materials Characterization and Engineering, 3 (2015) 237.
14. O. Jorquera, A. Kiperstok, E.A. Sales, M. Embirucu, M.L. Ghirardi, *Comparative energy life-cycle analyses of microalgal biomass production in open ponds and photobioreactors,* Bioresource Technology, 101 (2010) 1406–1413.

15. M.M. Jamel, P. Khoshnoud, S. Gunashekar, N. Abu-Zahra, *Effect of e-glass fibers and phlogopite mica on the mechanical properties and dimensional stability of rigid PVC foams*, Polymer-Plastics Technology and Engineering, 54 (2015) 1560–1570.
16. N.H. Abu-Zahra, M.M. Jamel, P. Khoshnoud, S. Gunashekar, *Enhanced dimensional stability of rigid PVC foams using glass fibers*, International Journal of Chemical, Nuclear, Materials and Metallurgical Engineering, 9 (2015) 12–17.
17. P. Khoshnoud, N. Abu-Zahra, *Properties of rigid polyvinyl chloride foam composites reinforced with different shape fillers*, Journal of Thermoplastic Composite Materials, (2016) 0892705716646417.
18. O. Pulz, *Photobioreactors: production systems for phototrophic microorganisms*, Applied Microbiology and Biotechnology, 57 (2001) 287–293.
19. B. Pendyala, M. Hanifzadeh, S. Viamajala, *Cultivation of low nitrogen algal biomass for high-quality algal biofuels production*, The 7th International Conference on Algal Biomass, Biofuels and Bioproducts, 2017.
20. B. Pendyala, M. Hanifzadeh, S. Viamajala, *Alkalphilic algal cultivation as a means for improved productivity and stability of algae-based production systems*, Algae Biomass Summit, 2016.
21. C.-Y. Chen, K.-L. Yeh, R. Aisyah, D.-J. Lee, J.-S. Chang, *Cultivation, photobioreactor design and harvesting of microalgae for biodiesel production: a critical review*, Bioresource Technology, 102 (2011) 71–81.
22. L. Christenson, R. Sims, *Production and harvesting of microalgae for wastewater treatment, biofuels, and bioproducts*, Biotechnology Advances, 29 (2011) 686–702.
23. M. Hanifzadeh, *Cultivation Strategies for Enhancement of Valuable Products Yield from Microalgae*, Science Alliance at the New York Academy of Sciences and PepsiCo, 2015.
24. F. Cucchiella, I. D'Adamo, M. Gastaldi, *Sustainable waste management: waste to energy plant as an alternative to landfill*, Energ Convers Manage, 131 (2017) 18–31.
25. V. Montemezzani, I.C. Duggan, I.D. Hogg, R.J. Craggs, *A review of potential methods for zooplankton control in wastewater treatment high rate algal ponds and algal production raceways*, Algal Research, 11 (2015) 211–226.
26. G. Vaseghi, A. Ghassemi, J. Loya, *Characterization of reverse osmosis and nanofiltration membranes: effects of operating conditions and specific ion rejection*, Desalination and Water Treatment, 57 (2016) 23461–23472.
27. G. Vaseghi, I. Celik, D. Apul, S. Burian, *Economic, Environmental, and Social Criteria Evaluation of Rainwater Harvesting System Options for an Office and Lab Building on the University of Utah Campus, Frontiers in Water Savings in Buildings*, Bentham Science Publishers, 2017, pp. 117–155.
28. L. Batan, J. Quinn, B. Willson, T. Bradley, *Net energy and greenhouse gas emission evaluation of biodiesel derived from microalgae*, Environmental Science & Technology, 44 (2010) 7975–7980.
29. A.F. Clarens, H. Nassau, E.P. Resurreccion, M.A. White, L.M. Colosi, *Environmental impacts of algae-derived biodiesel and bioelectricity for transportation*, Environmental Science & Technology, 45 (2011) 7554–7560.
30. H.H. Khoo, P.N. Sharratt, P. Das, R.K. Balasubramanian, P.K. Naraharisetti, S. Shaik, *Life cycle energy and CO_2 analysis of microalgae-to-biodiesel: preliminary results and comparisons*, Bioresource Technology, 102 (2011) 5800–5807.
31. J.K. Pittman, A.P. Dean, O. Osundeko, *The potential of sustainable algal biofuel production using wastewater resources*, Bioresource Technology, 102 (2010) 17–25.

32. R. Órpez, M.E. Martínez, G. Hodaifa, F. El Yousfi, N. Jbari, S. Sánchez, *Growth of the microalga Botryococcus braunii in secondarily treated sewage*, Desalination, 246 (2009) 625–630.
33. Y. Collos, P.J. Harrison, *Acclimation and toxicity of high ammonium concentrations to unicellular algae*, Marine Pollution Bulletin, 80 (2014) 8–23.
34. S. Cho, T.T. Luong, D. Lee, Y.K. Oh, T. Lee, *Reuse of effluent water from a municipal wastewater treatment plant in microalgae cultivation for biofuel production*, Bioresource Technology, 102 (2011) 8639–8645.
35. G. Markou, D. Georgakakis, *Cultivation of filamentous cyanobacteria (blue-green algae) in agro-industrial wastes and wastewaters: a review*, Applied Energy, 88 (2011) 3389–3401.
36. K.E. Dickinson, C.G. Whitney, P.J. McGinn, *Nutrient remediation rates in municipal wastewater and their effect on biochemical composition of the microalga Scenedesmus sp. AMDD*, Algal Research, 2 (2013) 127–134.
37. J.K. Pittman, A.P. Dean, O. Osundeko, *The potential of sustainable algal biofuel production using wastewater resources*, Bioresource Technology, 102 (2011) 17–25.
38. W. Mulbry, S. Kondrad, J. Buyer, *Treatment of dairy and swine manure effluents using freshwater algae: fatty acid content and composition of algal biomass at different manure loading rates*, Journal of Applied Phycology, 20 (2008) 1079–1085.
39. P. Chung, W. Pond, J. Kingsbury, E. Walker, L. Krook, *Production and nutritive value of Arthrospira platensis, a spiral blue-green alga grown on swine wastes*, Journal of Animal Science, 47 (1978) 319–330.
40. L. Wang, Y. Li, P. Chen, M. Min, Y. Chen, J. Zhu, R.R. Ruan, *Anaerobic digested dairy manure as a nutrient supplement for cultivation of oil-rich green microalgae Chlorella sp*, Bioresource Technology, 101 (2010) 2623–2628.
41. E.P. Lincoln, A.C. Wilkie, B.T. French, *Cyanobacterial process for renovating dairy wastewater*, Biomass and Bioenergy, 10 (1996) 63–68.
42. Q.-X. Kong, L. Li, B. Martinez, P. Chen, R. Ruan, *Culture of microalgae Chlamydomonas reinhardtii in wastewater for biomass feedstock production*, Applied Biochemistry and Biotechnology, 160 (2010) 9.
43. E. Olguin, S. Galicia, R. Camacho, G. Mercado, T. Pérez, *Production of Spirulina sp. in sea water supplemented with anaerobic effluents in outdoor raceways under temperate climatic conditions*, Applied Microbiology and Biotechnology, 48 (1997) 242–247.
44. S. Phang, M. Miah, B. Yeoh, M. Hashim, *Spirulina cultivation in digested sago starch factory wastewater*, Journal of Applied Phycology, 12 (2000) 395–400.
45. S. Chinnasamy, A. Bhatnagar, R.W. Hunt, K.C. Das, *Microalgae cultivation in a wastewater dominated by carpet mill effluents for biofuel applications*, Bioresource Technology, 101 (2010) 3097–3105.
46. M.R. Andrade, J.A.V. Costa, *Mixotrophic cultivation of microalga Spirulina platensis using molasses as organic substrate*, Aquaculture, 264 (2007) 130–134.
47. M. Hanifzadeh, S. Viamajala, *Effects of nutrients on microalgae cultivation*, Annual Graduate Research Symposium, 2014.
48. G. Markou, D. Vandamme, K. Muylaert, *Microalgal and cyanobacterial cultivation: the supply of nutrients*, Water Research, 65 (2014) 186–202.
49. B. Pendyala, M. Hanifzadeh, S. Viamajala, *Cultivation of low nitrogen algal biomass for high-quality algal biofuels production*, Algal Biomass Organization Summit, 2016.

50. D. Simionato, M.A. Block, N. La Rocca, J. Jouhet, E. Marechal, G. Finazzi, T. Morosinotto, *The response of nannochloropsis gaditana to nitrogen starvation includes de novo biosynthesis of triacylglycerols, a decrease of chloroplast galactolipids, and reorganization of the photosynthetic apparatus*, Eukaryotic Cell, 12 (2013) 665–676.
51. B. Pendyala, M. Hanifzadeh, S. Viamajala, *Enhanced biomass and lipid productivities of outdoor alkaliphillic microalgae cultures through increased media alkalinity*, AICHE 2016 Annual Meeting, 2016.
52. S.S. Merchant, J. Kropat, B. Liu, J. Shaw, J. Warakanont, *TAG, you're it! Chlamydomonas as a reference organism for understanding algal triacylglycerol accumulation*, Current Opinion Biotechnology, 23 (2012) 352–363.
53. K.-L. Yeh, J.-S. Chang, *Effects of cultivation conditions and media composition on cell growth and lipid productivity of indigenous microalga Chlorella vulgaris ESP-31*, Bioresource Technology, 105 (2012) 120–127.
54. Z. Nabati, M. Hanifzadeh, E.C. Garcia, S. Viamajala, *Investigation of Chlorella sp cultivation in low concentration of Mg and Ca*, 38th Symposium on Biotechnology for Fuels and Chemicals, 2016.
55. X.D. Deng, X.W. Fei, Y.J. Li, *The effects of nutritional restriction on neutral lipid accumulation in Chlamydomonas and Chlorella*, African Journal of Microbiology Research, 5 (2011) 260–270.
56. E.C. Garcia, M. Hanifzadeh, S. Viamajala, *The effect of Bicarbonate on microalgae*, Midwest Graduate Research Symposium, 2015.
57. A. Juneja, R.M. Ceballos, G.S. Murthy, *Effects of environmental factors and nutrient availability on the biochemical composition of algae for biofuels production: a review*, Energies (Basel, Switz.), 6 (2013) 4607–4638, 4632.
58. S. Zhu, W. Huang, J. Xu, Z. Wang, J. Xu, Z. Yuan, *Metabolic changes of starch and lipid triggered by nitrogen starvation in the microalga Chlorella zofingiensis*, Bioresource Technology, 152 (2014) 292–298.
59. B. Fernandes, J. Teixeira, G. Dragone, A.A. Vicente, S. Kawano, K. Bišová, P. Přibyl, V. Zachleder, M. Vítová, *Relationship between starch and lipid accumulation induced by nutrient depletion and replenishment in the microalga Parachlorella kessleri*, Bioresource Technology, 144 (2013) 268–274.
60. Y. Tan, J. Lin, *Biomass production and fatty acid profile of a Scenedesmus rubescens-like microalga*, Bioresource Technology, 102 (2011) 10131–10135.
61. S. Venkata Mohan, M.P. Devi, *Salinity stress induced lipid synthesis to harness biodiesel during dual mode cultivation of mixotrophic microalgae*, Bioresource Technology, 165 (2014) 288–294.
62. S.-H. Ho, J.-S. Chang, Y.-Y. Lai, C.-N.N. Chen, *Achieving high lipid productivity of a thermotolerant microalga Desmodesmus sp. F2 by optimizing environmental factors and nutrient conditions*, Bioresource Technology, 156 (2014) 108–116.
63. Q. He, H. Yang, L. Xu, L. Xia, C. Hu, *Sufficient utilization of natural fluctuating light intensity is an effective approach of promoting lipid productivity in oleaginous microalgal cultivation outdoors*, Bioresource Technology, 180 (2015) 79–87.
64. K. Sharma, Y. Li, P.M. Schenk, *UV-C-mediated lipid induction and settling, a step change towards economical microalgal biodiesel production*, Green Chemistry, 16 (2014) 3539–3548.
65. G. Procházková, I. Brányiková, V. Zachleder, T. Brányik, *Effect of nutrient supply status on biomass composition of eukaryotic green microalgae*, Journal of Applied Phycology, 26 (2014) 1359–1377.

66. J. Msanne, D. Xu, A.R. Konda, J.A. Casas-Mollano, T. Awada, E.B. Cahoon, H. Cerutti, *Metabolic and gene expression changes triggered by nitrogen deprivation in the photoautotrophically grown microalgae Chlamydornonas reinhardtii and Coccomyxa sp C-169*, Phytochemistry, 75 (2012) 50–59.
67. A.J. Klok, D.E. Martens, R.H. Wijffels, P.P. Lamers, *Simultaneous growth and neutral lipid accumulation in microalgae*, Bioresource Technology, 134 (2013) 233–243.
68. A.I. Barros, A.L. Gonçalves, M. Simões, J.C. Pires, *Harvesting techniques applied to microalgae: a review*, Renewable and Sustainable Energy Reviews, 41 (2015) 1489–1500.
69. J.J. Milledge, S. Heaven, *A review of the harvesting of micro-algae for biofuel production*, Reviews in Environmental Science and Bio/ Technology, 12 (2013) 165–178.
70. T. Chatsungnoen, Y. Chisti, *Harvesting microalgae by flocculation–sedimentation*, Algal Research, 13 (2016) 271–283.
71. A.J. Garzon-Sanabria, S.S. Ramirez-Caballero, F.E. Moss, Z.L. Nikolov, *Effect of algogenic organic matter (AOM) and sodium chloride on Nannochloropsis salina flocculation efficiency*, Bioresource Technology, 143 (2013) 231–237.
72. F. Roselet, D. Vandamme, M. Roselet, K. Muylaert, P.C. Abreu, *Effects of pH, salinity, biomass concentration, and algal organic matter on flocculant efficiency of synthetic versus natural polymers for harvesting microalgae biomass*, BioEnergy Research, 10 (2017) 427–437.
73. A. Singh, P.S. Nigam, J.D. Murphy, *Mechanism and challenges in commercialisation of algal biofuels*, Bioresource Technology, 102 (2010) 26–34.
74. K.-Y. Show, D.-J. Lee, A.S. Mujumdar, *Advances and challenges on algae harvesting and drying*, Drying Technology, 33 (2015) 386–394.
75. N. Uduman, Y. Qi, M.K. Danquah, G.M. Forde, A. Hoadley, *Dewatering of microalgal cultures: a major bottleneck to algae-based fuels*, Journal of Renewable and Sustainable Energy, 2 (2010) 012701.
76. J. Zhang, B. Hu, *A novel method to harvest microalgae via co-culture of filamentous fungi to form cell pellets*, Bioresource Technology, 114 (2012) 529–535.
77. S.H. Abdul Hamid, F. Lananan, W.N.S. Din, S.S. Lam, H. Khatoon, A. Endut, A. Jusoh, *Harvesting microalgae, Chlorella sp. by bio-flocculation of Moringa oleifera seed derivatives from aquaculture wastewater phytoremediation*, International Biodeterioration & Biodegradation, 95 (2014) 270–275.
78. W. Zhou, M. Min, B. Hu, X. Ma, Y. Liu, Q. Wang, J. Shi, P. Chen, R. Ruan, *Filamentous fungi assisted bio-flocculation: a novel alternative technique for harvesting heterotrophic and autotrophic microalgal cells*, Separation and Purification Technology, 107 (2013) 158–165.
79. K.A. Al-Hothaly, E.M. Adetutu, M. Taha, D. Fabbri, C. Lorenzetti, R. Conti, B. H. May, S.S. Shar, R.A. Bayoumi, A.S. Ball, *Bio-harvesting and pyrolysis of the microalgae Botryococcus braunii*, Bioresource Technology, 191 (2015) 117–123.
80. C. Banerjee, S. Ghosh, G. Sen, S. Mishra, P. Shukla, R. Bandopadhyay, *Study of algal biomass harvesting through cationic cassia gum, a natural plant based biopolymer*, Bioresource Technology, 151 (2014) 6–11.
81. M. Castrillo, L.M. Lucas-Salas, C. Rodríguez-Gil, D. Martínez, *High pH-induced flocculation–sedimentation and effect of supernatant reuse on growth rate and lipid productivity of Scenedesmus obliquus and Chlorella vulgaris*, Bioresource Technology, 128 (2013) 324–329.

82. C. Yoo, H.J. La, S.C. Kim, H.M. Oh, *Simple processes for optimized growth and harvest of Ettlia sp. by pH control using CO2 and light irradiation*, Biotechnology and Bioengineering, 112 (2015) 288–296.
83. Z. Wu, Y. Zhu, W. Huang, C. Zhang, T. Li, Y. Zhang, A. Li, *Evaluation of flocculation induced by pH increase for harvesting microalgae and reuse of flocculated medium*, Bioresource Technology, 110 (2012) 496–502.
84. K. Spilling, J. Seppälä, T. Tamminen, *Inducing autoflocculation in the diatom Phaeodactylum tricornutum through CO2 regulation*, Journal of Applied Phycology, 23 (2011) 959–966.
85. S. Salim, M.H. Vermuë, R.H. Wijffels *Wijffels, Ratio between autoflocculating and target microalgae affects the energy-efficient harvesting by bio-flocculation*, Bioresource Technology, 118 (2012) 49–55.
86. S. Salim, Z. Shi, M.H. Vermuë, R.H. Wijffels, *Effect of growth phase on harvesting characteristics, autoflocculation and lipid content of Ettlia texensis for microalgal biodiesel production*, Bioresource Technology, 138 (2013) 214–221.
87. Y. Shen, Z. Fan, C. Chen, X. Xu, *An auto-flocculation strategy for Chlorella vulgaris*, Biotechnol Lett, 37 (2015) 75–80.
88. Q. Hu, M. Sommerfeld, E. Jarvis, M. Ghirardi, M. Posewitz, M. Seibert, A. Darzins, *Microalgal triacylglycerols as feedstocks for biofuel production: perspectives and advances*, The Plant Journal, 54 (2008) 621–639.
89. K.-Y. Show, D.-J. Lee, J.-H. Tay, T.-M. Lee, J.-S. Chang, *Microalgal drying and cell disruption–recent advances*, Bioresource Technology, 184 (2015) 258–266.
90. M.A. Borowitzka, *High-value products from microalgae-their development and commercialisation*, Journal of Applied Phycology, 25 (2013) 743–756.
91. M. Aziz, T. Oda, T. Kashiwagi, *Integration of energy-efficient drying in microalgae utilization based on enhanced process integration*, Energy, 70 (2014) 307–316.
92. A. Guldhe, B. Singh, I. Rawat, K. Ramluckan, F. Bux, *Efficacy of drying and cell disruption techniques on lipid recovery from microalgae for biodiesel production*, Fuel, 128 (2014) 46–52.
93. Y. Chisti, *Biodiesel from microalgae*, Biotechnology Advances, 25 (2007) 294–306.
94. A. Karmakar, S. Karmakar, S. Mukherjee, *Properties of various plants and animals feedstocks for biodiesel production*, Bioresource Technology, 101 (2010) 7201–7210.
95. H. El-Shimi, N.K. Attia, S. El-Sheltawy, G. El-Diwani, *Biodiesel production from Spirulina-platensis microalgae by in-situ transesterification process*, Journal of Sustainable Bioenergy Systems, 3 (2013) 224.
96. D.R. Vardon, B.K. Sharma, G.V. Blazina, K. Rajagopalan, T.J. Strathmann, *Thermochemical conversion of raw and defatted algal biomass via hydrothermal liquefaction and slow pyrolysis*, Bioresource Technology, 109 (2012) 178–187.
97. S. Edmundson, M. Huesemann, R. Kruk, T. Lemmon, J. Billing, A. Schmidt, D. Anderson, *Phosphorus and nitrogen recycle following algal bio-crude production via continuous hydrothermal liquefaction*, Algal Research, 26 (2017) 415–421.
98. Y. Guo, T. Yeh, W. Song, D. Xu, S. Wang, *A review of bio-oil production from hydrothermal liquefaction of algae*, Renewable and Sustainable Energy Reviews, 48 (2015) 776–790.
99. C. Tian, B. Li, Z. Liu, Y. Zhang, H. Lu, *Hydrothermal liquefaction for algal biorefinery: a critical review*, Renewable and Sustainable Energy Reviews, 38 (2014) 933–950.

100. D.L. Barreiro, W. Prins, F. Ronsse, W. Brilman, *Hydrothermal liquefaction (HTL) of microalgae for biofuel production: state of the art review and future prospects,* Biomass and Bioenergy, 53 (2013) 113–127.
101. D. Tang, W. Han, P. Li, X. Miao, J. Zhong, *CO_2 biofixation and fatty acid composition of Scenedesmus obliquus and Chlorella pyrenoidosa in response to different CO_2 levels,* Bioresource Technology, 102 (2011) 3071–3076.
102. S. Singhasuwan, W. Choorit, S. Sirisansaneeyakul, N. Kokkaew, Y. Chisti, *Carbon-to-nitrogen ratio affects the biomass composition and the fatty acid profile of heterotrophically grown Chlorella sp.* TISTR 8990 for biodiesel production, Journal of Biotechnology, 216 (2015) 169–177.
103. L. Borges, S. Caldas, M.G. Montes D'Oca, P.C. Abreu, *Effect of harvesting processes on the lipid yield and fatty acid profile of the marine microalga Nannochloropsis oculata,* Aquaculture Reports, 4 (2016) 164–168.
104. M. Mofijur, H. Masjuki, M. Kalam, A. Atabani, M. Shahabuddin, S. Palash, M. Hazrat, *Effect of biodiesel from various feedstocks on combustion characteristics, engine durability and materials compatibility: a review,* Renewable and Sustainable Energy Reviews, 28 (2013) 441–455.
105. D.-S. Kim, M. Hanifzadeh, A. Kumar, *Trend of biodiesel feedstock and its impact on biodiesel emission characteristics, Environmental Progress & Sustainable Energy,* n/a-n/a.
106. R.M. Joshi, M.J. Pegg, *Flow properties of biodiesel fuel blends at low temperatures,* Fuel, 86 (2007) 143–151.
107. M.J. Ramos, C.M. Fernández, A. Casas, L. Rodríguez, Á. Pérez, *Influence of fatty acid composition of raw materials on biodiesel properties,* Bioresource Technology, 100 (2009) 261–268.
108. J.K. Bwapwa, A. Anandraj, C. Trois, *Possibilities for conversion of microalgae oil into aviation fuel: a review,* Renewable and Sustainable Energy Reviews, 80 (2017) 1345–1354.
109. G. Göttlicher, P. Schlagermann, R. Dillschneider, R. Rosello-Sastre, C. Posten, *Composition of algal oil and its potential as biofuel,* Journal of Combustion, 2012 (2012) 14.
110. Y.-H. Chen, B.-Y. Huang, T.-H. Chiang, T.-C. Tang, *Fuel properties of microalgae (Chlorella protothecoides) oil biodiesel and its blends with petroleum diesel,* Fuel, 94 (2012) 270–273.
111. S.K. Mandotra, P. Kumar, M.R. Suseela, S. Nayaka, P.W. Ramteke, *Evaluation of fatty acid profile and biodiesel properties of microalga Scenedesmus abundans under the influence of phosphorus, pH and light intensities,* Bioresource Technology, 201 (2016) 222–229.
112. S. Ruangsomboon, *Effects of different media and nitrogen sources and levels on growth and lipid of green microalga Botryococcus braunii KMITL and its biodiesel properties based on fatty acid composition,* Bioresource Technology, 191 (2015) 377–384.
113. G. Tüccar, T. Özgür, K. Aydın, *Effect of diesel–microalgae biodiesel–butanol blends on performance and emissions of diesel engine,* Fuel, 132 (2014) 47–52.
114. V. Makarevičienė, S. Lebedevas, P. Rapalis, M. Gumbyte, V. Skorupskaite, J. Žaglinskis, *Performance and emission characteristics of diesel fuel containing microalgae oil methyl esters,* Fuel, 120 (2014) 233–239.

6

Biodiesel Properties Depending on Blends and Feedstocks

Cloud Point, Kinematic Viscosity, and Flash Point

Sudheer Kumar Kuppili, Ashok Kumar, and Dong-Shik Kim

Abbreviations

FAME – Fatty Acid Methyl Esters
ULSD – Ultra Low Sulfur Diesel
SME – Soybean Methyl Esters
TO – Tallow Oil
WCO – Waste Cooking Oil

6.1 Introduction

Over the past few decades, fossil fuel energy consumption has increased at an expeditious pace, which leaves challenging targets to meet the high energy demand [1]. To satisfy the ever increasing energy demand, the recovery of fossil fuel has gone up, and as a result, scarcity of conventional fossil fuels has increased. At the same time, interest in the usage of alternative energy sources due to the recent hike in the prices of petroleum has increased as well as environmental concerns [2–4]. Biodiesel is one of the most readily available renewable sources that can be used without too much hassle in the current infrastructure (US EIA, 2012). Also because of its innate characteristics biodiesel can be used in any diesel vehicle without performing any major modifications (Omidvarborna et al., 2016). It has become more attractive recently because of its greater environmental benefits than its par and also of its sustainability [5]. Its feedstocks are carbon neutral and have low sulfur content that contributes greatly in the mitigation of environmental effects such as climate change and air pollution [6].

Biodiesel can be generated from various feedstocks such as vegetable oil, animal fat, grease, waste cooking oil, algae, and so on and its feedstock is becoming diversified rapidly [7]. Transesterification process is commonly used with an alcohol (methanol or ethanol) and an acid or base catalyst. Because of its similar properties to petroleum diesel, biodiesel can be put to use in vehicles run on diesel. In the United States, B20 diesel (biodiesel

20%, ULSD 80%) is commonly used in transportation, while 10% (B10), 50% (B50), and 100% (B100) biodiesel fuels are available. Prior to its use, it is important to know the fuel properties of biodiesel. To completely replace petroleum-based diesel or to use biodiesel blends for diesel engines, the accurate information of physical properties of biodiesel is needed to meet the engine conditions. The property information is critical in deciding the type of additives, engine operating conditions, requirements for vehicle survival in harsh climates. The information commonly includes, but not limited to, cloud point, pour point, kinematic viscosity, cetane number, flash point, ash content, sulfur content, carbon residue, acid value, copper corrosion, and higher heating value [3,4,8].

In this study, three different commercial biodiesels and its various blends were tested for cloud point, kinematic viscosity, and flash point. Biodiesels from tallow oil, soybean, and waste cooking oil were selected to test. The results were compared with ULSD fuel. Characteristics and test methods are summarized as follows:

Cloud Point: For petroleum products and biodiesel fuels, cloud point is an index of the lowest temperature of their utility for certain applications [9]. Below a cloud point fuel tends to solidify and form crystals, which restrains the movement of engine parts and alters its working efficiency [3]. The ASTM D2500 procedure was used to test the cloud point.

Kinematic Viscosity: Petroleum and non-petroleum products are used as lubricants depending on the viscosity of the fuel being used in an engine. Therefore, it is important to know viscosity of a fuel to select a right lubricant to determine a proper amount of it. Moreover fuel management plan for storage, handling and setting operational conditions should consider viscosity [10]. Viscosity of the vegetable oils is 9 to 17 times greater than the petroleum diesel fuel [11]. Kinematic viscosity is the coefficient of viscosity of a fluid divided by its density, usually measured in centistokes (cSt). ASTM D445 standards were used for this test.

Flash Point: The flash point measures the specimen's tendency to form a flammable mixture with air under controlled laboratory conditions [12]. It is the only one among the properties that should be considered in estimating the overall flammability hazard of a material [12,13]. Basic objective of flash point measurement is to limit the amount of alcohol in biofuel [14]. The abnormality of the flash point can be associated with contamination of the material. ASTM limit for biodiesel flash point is 130°C and for diesel it is 38°C [12]. ASTM D92-Cleveland open cup test standards were used.

6.2 Material and Methods

First cloud point, kinematic viscosity, and flash point tests were conducted on ULSD (i.e., B0) as a control. Then, B10, B20, B50, and B100 blends of

TABLE 6.1

Bath and sample temperature ranges [9]

Bath	Bath Temperature Setting (°C)	Sample Cloud Point Temperature Range (°C)
1	0 ± 1.5	Start to 9
2	-18 ± 1.5	9 to -6
3	-33 ± 1.5	-6 to -24
4	-51 ± 1.5	-24 to -42
5	-69 ± 1.5	-42 to -60

soybean (SME), tallow oil (TO), and waste cooking oil (WCO) biodiesel were tested.

6.2.1 Cloud Point

A water bath was used for the cloud point test. A test jar of 32.5 mm diameter and cork with a provision to insert thermometer was used to contain a test sample. A 40 mm layer of a test sample was transferred into the test jar. According to the ASTM D2500 standards, different water baths were prepared using sodium chloride (NaCl), calcium chloride ($CaCl_2$), dry ice, water, and ice. The test fuel was placed inside the bath and checked for a cloud point and if it didn't happen in one bath, then another bath of different temperature range was used and a cloud point temperature was recorded (ASTM D2500-02, 2002). Two pre-calibrated thermometers, that is, a Digi-Sense glass thermometer and a DURAC plus low cloud and pour glass thermometer (ASTM No. 6C) were used which has a range from –35°C to 50°C (fuel sample temperature measurement) and -80°C to 20°C (bath temperature measurement), respectively, with an accuracy of ±1.5°C. The thermometer measuring the fuel sample temperature must be placed in such a way that it touches bottom surface of the test jar and the thermometer measuring the bath temperature was immersed to a depth of 135–125 mm from the surface. The placement and immersion depth of thermometer in the test jar are important in this test because cloud point is observed at the bottom of the test jar. So, the exact temperature can be measured only by placing the thermometer to touch the bottom surface. The thermometer measuring the bath temperature must be at the specified depth because it is the depth at which test jar is placed. Therefore, exact temperature of the bath which is in contact with the test jar can be measured. Table 6.1 shows the bath and sample temperature ranges. Test was repeated three times to check the consistency of the results.

6.2.2 Kinematic Viscosity

Kinematic viscosity test is an essential part of the tests because viscosity directly affects engine performance. As it controls atomization of a fuel upon injection into a combustion chamber, it may ultimately cause formation of engine deposits that reduces engine performance. The higher is the viscosity, the greater is the tendency of fuel to cause such problems. For the kinematic viscosity test, a calibrated viscometer was used. The test sample was kept in a beaker maintained at 40°C. ASTM D445 standards were used. The temperature was measured using a pre-calibrated Digi-Sense glass thermometer which has a range from -35°C to 50°C with an accuracy of ±1.5°C. Briefly a calibrated Cannon-Fenske Routine Viscometer, size 75, was filled with 7 ml of test sample and placed inside a 40°C beaker for 10 minutes. Then the meniscus was pulled up to the upper marked line and released. A stop watch was used to measure the time taken by the sample to pass the lower marked circle on the viscometer (ASTM D445-70, 1970). The test was repeated several times to confirm the consistency of the results.

6.2.3 Flash Point

Flash point test is to measure the lowest temperature at which an ignitable vapor is produced under atmospheric pressure due to heating of petroleum products or biodiesel [12]. To ensure the safety with a stored flammable liquid, it is obligator to know the flash point of it [13]. To carry out the flash point test, a Cleveland open test cup was used. Test flame was generated by using a kitchen lighter and was setup at low flame. Heating plate, thermometer and cup holders were used. A sample of 70 ml was taken into the Cleveland open test cup and placed over a heating plate. According to the ASTM method, it is heated at a rate of 5–17°C/min (here, we used 10±1°C/min) and the rate of temperature increase was reduced to 5 to 6°C for the last 28°C before the flash point. The temperature rise was measured using a pre-calibrated Thermco glass thermometer which can measure from -10°C to 210°C with an accuracy of ±1°C. Thermometer was placed in the biodiesel in such a way that the mercury bulb completely immersed into the fuel but it didn't touch the bottom surface of the cup. A test flame was passed over the liquid to ignite the possible existing flammable vapors and the temperature at which a blue color flame appears for less than a second was noted as flash point (ASTM D92-05, 2005). If the thermometer immersed in the fuel touches the bottom surface of the cup, the flash point temperature measured might increase by few degrees Celsius.

6.2.4 Uncertainties in Measurements

In this study uncertainty in the experimental values can be expected during the measurement of temperature using a thermometer and time with a stopwatch. The major sources of human error experienced in thermometer measurement were parallax error and position of the thermometer in the bath solutions and fuel samples. We reduced the thermometer measurement resolution reading error by using a magnifying glass. Multiple experiments were performed to obtain standard deviations of the data points within a 95% confidence interval.

6.3 Results and Discussion

6.3.1 Cloud Point

Cloud point of biodiesel is reported to vary with the type of feedstock, ester composition and the presence of additives [11,15]. Manufacturing companies adjusts the cloud point of a particular biodiesel depending on the season and location. The cloud point of a fuel in summer may be adjusted higher than in winter or any other season in comparison. On the other hand, the cloud point in temperate or cold regions should be lower than that in tropical regions. Furthermore, as biodiesel properties depends on a feedstock, biodiesel obtained from semi-drying oils like rapeseed and soybean has better cold flow properties than biodiesel obtained from animal fat such as beef and pig tallow [16]. This trend was also testified in the current result as shown in Table 6.2 and depicted in the form of linear line in Figures 6.1a–c. Error bars represent a 95% confidence interval. The cloud point of soybean biodiesel (B100), that is, SME, was measured

TABLE 6.2

Cloud point temperatures of ULSD and its biodiesel blends

Fuel	Blend	Cloud Point Obtained (°C)
ULSD	-	-9.000 ± 0.500
White grease biodiesel (Tallow oil)	B100	9.000 ± 0.478
	B50	0.000 ± 0.000
	B20	-5.000 ± 0.408
	B10	-6.500 ± 0.250
Soybean biodiesel (SME)	B100	0.000 ± 0.500
	B50	-4.500 ± 0.250
	B20	-7.000± 0.288
	B10	-7.500± 0.250
Waste cooking oil biodiesel (WCO)	B100	3.000± 0.250
	B50	-4.500± 0.288
	B20	-6.500± 0.250
	B10	-8.000± 0.707

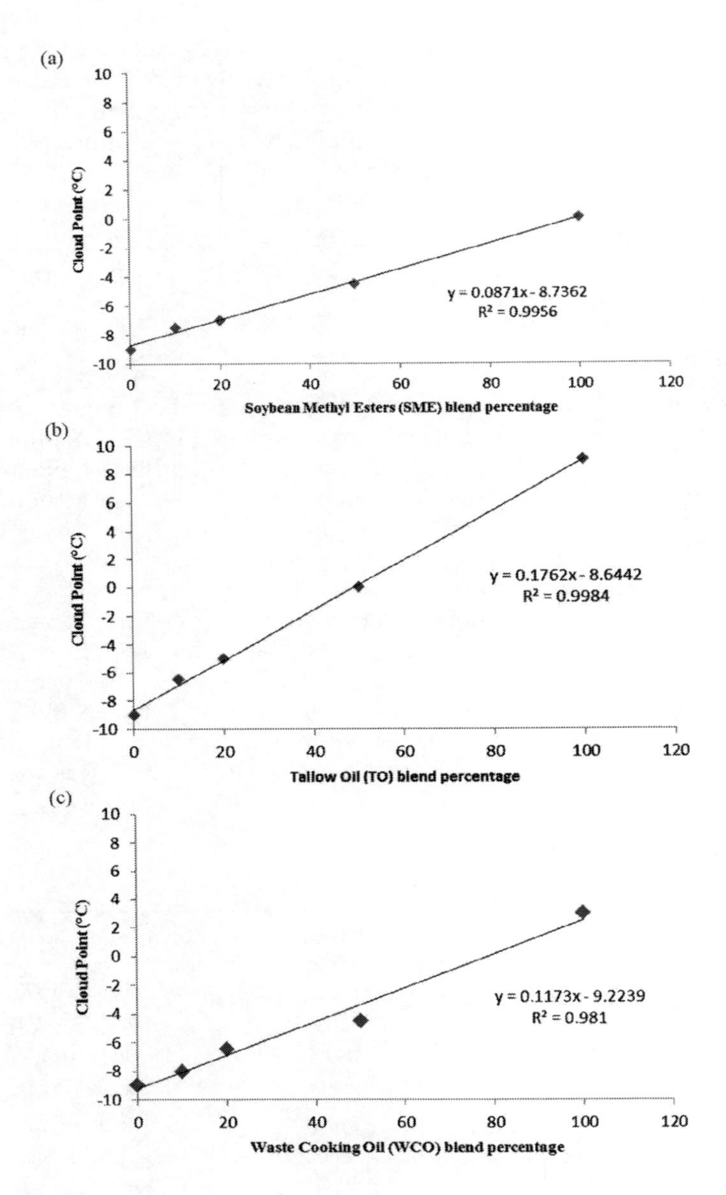

FIGURE 6.1
Linear regressions for cloud points of (a) soybean (SME) biodiesel, (b) tallow oil (TO) biodiesel and (c) waste cooking oil (WCO) biodiesel and their blends.

0.0°C and that of waste cooking oil and tallow oil (TO) biodiesel (B100) were 3.0°C and 9.0°C, respectively. Regardless of the feedstock type, B100 biodiesels have higher cloud points (9.0, 0.0, 3.0°C) than ULSD (-9.0°C).

Table 6.2 shows the effect of blending on cloud points. For the three types of biodiesels tested in this study, their cloud points varied differently for the biodiesel content. As shown in Figure 6.1a, the increase of soybean biodiesel content raised the cloud point from -7.5°C to 0.0°C. The cloud point increasing rate was 0.087°C/blend percent in terms of biodiesel blending, which means the cloud point of soybean biodiesel increases by average 0.8°C for a 10% increase of its content in the blends.

For TO biodiesel, the dependence of cloud point on blending is 0.176 as shown in Figure 6.1b. This means that the cloud point of TO biodiesel increases by average 1.7°C for a 10% increase in its content in the blends. This increase rate is about 2 times greater than that of SME biodiesel in Figure 6.1a. Therefore, TO biodiesel raises the cloud point more than SME biodiesel when blending with ULSD.

The biodiesel blends made of waste cooking oil (WCO) also show a linear increase in terms of the blending concentration of biodiesel similar to tallow oil and soybean biodiesels. The dependence of cloud point on blending is calculated 0.117°C/blending percent (the slope between B0 and B100), which is higher than SME biodiesel and lower than TO biodiesel (Figure 6.1c). The reliability and accuracy of the rate of increase of cloud point temperature with respect to the percentage of biodiesel content can be validated by the R^2 value that is displayed on Figures 6.1a-c. It is 0.9956, 0.9984 and 0.981 for SME, TO, and WCO, respectively.

There is no particular range specified for cloud point of petroleum fuels and biodiesels in the United States of America due to the considerable climate variation in the nation. Various research studies show similar cloud point values for SME, TO and WCO. From the experimental results of various authors, the cloud point of neat soybean biodiesel (B100 of SME) was reported in a temperature range of 2 to -5°C [3,16–18]. Chiu et al. [3] measured the cloud point of neat SME biodiesel (B100) as -4°C, which was 4°C lower than what we have measured, while Lee et al. [19] reported the cloud point as -2°C. Chui et al. [3] also measured the cloud point for a series SME blends (B0, B20, B30 and B40) with low sulfur diesel (LSD), and the results varied from -12°C for B40 to -18°C for B0. But, in the current research ULSD was used and a different series of SME blends (B0, B10, B20, B50 and B100) were tested. Wyatt et al. [18] measured cloud point of tallow oil biodiesel (B100) and it is 11°C, which is 2°C more than the current study results, while Canakci and Sanli [7] recorded a value of 12°C. On the other hand, Encinar et al. [20] experimented with operational parameters such as methanol/waste cooking oil ratio, catalyst, and catalyst percentage, and measured cloud point for all those cases, which ranged from 1.1–6.1°C. Meanwhile, Chhetri et al. [21] recorded -1°C in their experiments on WCO (B100), which is 4°C less than this current study's measured value for WCO (B100) that is, 3°C.

In general, feedstocks with a high percentage of saturated fatty acid have high cloud point and pour point, which are regarded as bad cold flow properties [22,23]. SME has less amount saturated fatty acids than TO and WCO [22]. In other words, SME has better cold flow properties because of higher linoleic (C18:2) and linolenic (C18:3) non-saturated fatty acid contents than TO and WCO. WCO is richer in stearic acid (C18:0, saturated) and oleic acid (C18:1), whereas 90% of TO is filled with saturated palmitic (C16:0), stearic (C18:0), and partially with saturated oleic (C18:1) fatty acids [18,24]. The very reason for obtaining different cloud point values in this current research from the other reported data is the existence of a different fatty acids profile. The chemical structures and composition of the fatty acids in the biodiesel samples used for this research were analyzed in other studies using Fourier Transform Infrared Spectroscopy (FTIR) [25]. It was revealed that SME has comparatively more unsaturated carbon chains than WCO and TO.

As shown in Table 6.2 and Figures 6.1a–c, the results of the cloud point experiment indicate the cloud point of SME and its blends B10, B20, and B50 have relatively higher values than those of TO-based and WCO-based biodiesels. Therefore SME-based biodiesel helped a biodiesel blend have better cold flow properties than the other feedstock-based biodiesels with the same blending. TO has higher cloud point than the remaining two biodiesels and invariably its blends with ULSD also possess high cloud point values. To lower the cloud point temperature of commercial biodiesels, they were blended with ULSD, which has a cloud point temperature of –9°C. Among the blends, B10 of SME has the lowest cloud point, that is, -7.5°C with standard deviation of ± 0.25°C, whereas B10 of TO has the lowest cloud point, -6.5°C. Furthermore, it is clear that SME's cloud point tends to change less with the ULSD content than TO and WCO when it is blended with ULSD. From the slope equations of the linearity lines (displayed on Figures 6.1a–c) it can be understood that SME's cloud point tends to change by a lesser value, than TO and WCO when it is blended with ULSD.

Figure 6.2 indicates that the existence of long carbon chains of saturated fatty acids is another possible reason for high cloud points which means bad cold flow properties of TO and WCO B100 biodiesels. Long carbon chains and high molecular weights have been reported to be contributors to the high cloud points of biodiesel [23,24]. Again, characteristic effects of fatty acid composition on cold flow properties are depicted in Figure 6.2, which summarizes that the biodiesel with high compositions of unsaturated fatty acids, branched and short chain fatty esters exhibits improved cold flow properties [23]. Branched carbon chain fatty acids may easily form a structure which needs high thermodynamic energy for crystallization of fatty acid chains. Due to this effect, biodiesel doesn't get solidified easily which means better cold flow properties [19,23,26,27].

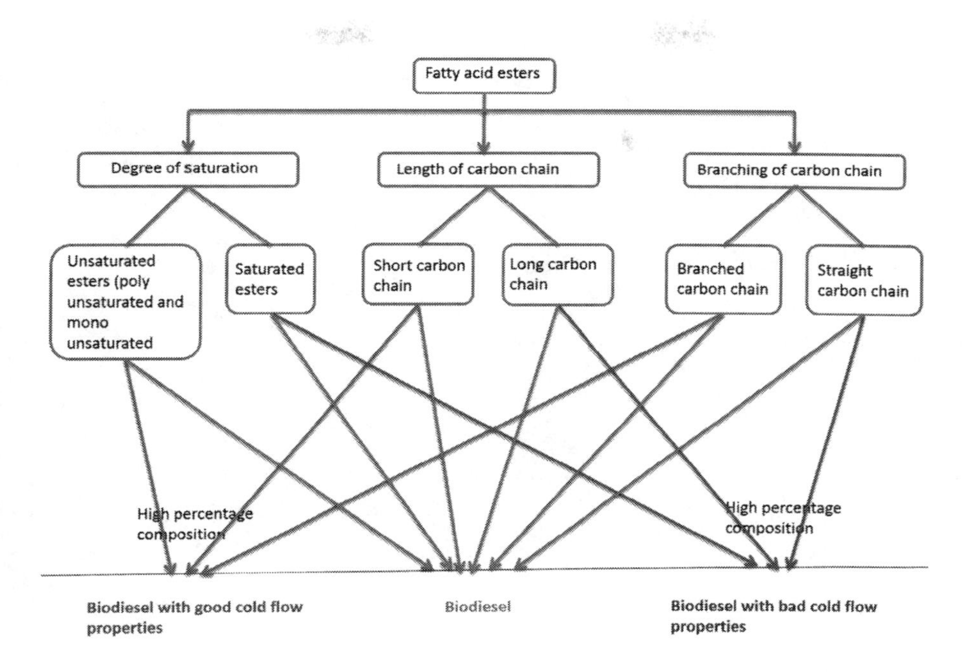

FIGURE 6.2
Effects of fatty acid composition on cold flow properties.

A few data points deviate from the linear fitting lines in Figures 6.1a–c, are thought to be due to human errors during the making of biodiesel blends, equivocal decision of a cloud formation point at the bottom surface of fuel by observation, and inconsistent reading of a cloud point temperature on the thermometer.

6.3.2 Kinematic Viscosity

High viscosity is the major problem with most of the biodiesels produced from feedstocks such as vegetable oil, animal fat, and waste oil [11,28]. Because of this reason, B100 of any biodiesel has been rarely used, and blends with ULSD are mostly used. ASTM has proposed a range of kinematic viscosity of fuel, where fuel is safe to use in a vehicle and is free of clogging, or arresting the engine parts movement. The suggested kinematic viscosity range for biodiesel is 1.9–6 cSt and 1.9–4.1 cSt for ULSD [10].

The current test was repeated several times in order to get the statistical confidence interval of the data. The precision of the test methods and results were determined using the statistical examination of the laboratory results [9]. The test results are summarized in Table 6.3 and demonstrated

TABLE 6.3

Kinematic viscosities of ULSD and its biodiesel blends

Fuel	Blend	Kinematic Viscosity at 40°C (cSt), ν = C * t
ULSD	-	3.439 ± 0.031
Soybean biodiesel	B100	4.912 ± 0.013
	B50	3.884 ± 0.026
	B20	3.564 ± 0.009
	B10	3.383 ± 0.018
White grease biodiesel (TALLOW)	B100	5.633 ± 0.030
	B50	4.219 ± 0.013
	B20	3.693 ± 0.033
	B10	3.426 ± 0.012
Waste cooking oil biodiesel	B100	5.640 ± 0.021
	B50	4.282 ± 0.007
	B20	3.699 ± 0.007
	B10	3.507 ± 0.021

in Figure 6.3. As expected, the kinematic viscosities of 100% TO and WCO biodiesels (5.63 and 5.64 cSt) are much higher than that of ULSD (3.55 cSt). SME biodiesel showed 4.91 cSt, which is lower than TO and WCO biodiesels, but about 1.4 times greater than that of ULSD. As the content of ULSD increases in blends, B50 (ULSD 50%), B20 (ULSD 80%), B10 (ULSD 90%), kinematic viscosity decreases.

As shown in Figures 6.3a–c, kinematic viscosity increases almost linearly as the blending ratio of biodiesel increases. The increasing rate of kinematic viscosity for SME biodiesel is 0.0153 cSt/BD percent, a slope between B0 and B100. For TO biodiesel, the slope is 0.0226 cSt/BD percent, and 0.0225 cSt/BD percent for WCO biodiesel. This means that the kinematic viscosity increased by 0.153 cSt for every 10% SME biodiesel blending, and approximately 0.23 cSt for both TO and WCO biodiesels. Therefore, for the same biodiesel contents, TO and WCO showed a little higher kinematic viscosity than that of SME by 0.1–0.3 cSt, and they showed about 50% greater dependence on blending than SME. Compared to the cloud points of blends, biodiesel content appears to control the cloud points of blends more significantly than viscosity of blends.

Kinematic viscosity measurement helps selecting the profile of fatty acids required in the feedstock that is used for biodiesel production [29]. Viscosities of methyl or ethyl esters of vegetable oil are nearly twice that of diesel fuels [30]. From Table 6.3, kinematic viscosity of soybean oil biodiesel is nearly 1.43 times that of ULSD, while tallow oil and waste cooking oil has 1.65 times higher than ULSD. The longer is the length of the fatty

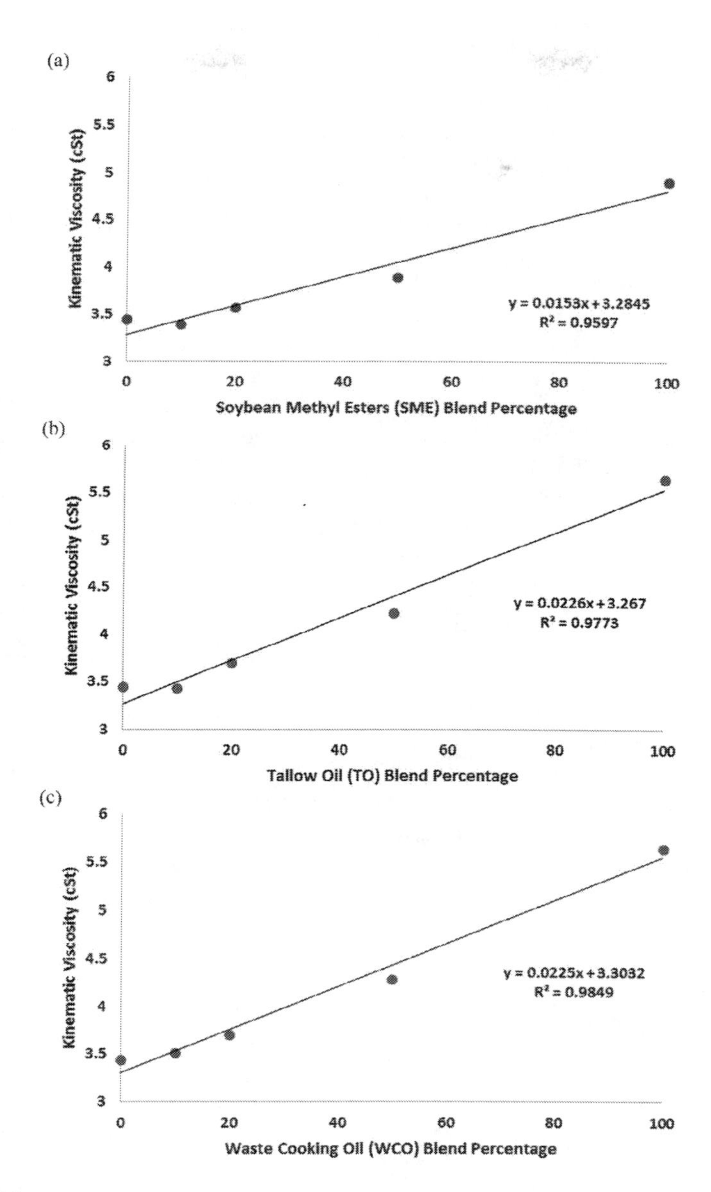

FIGURE 6.3
Linear regressions for kinematic viscosities of (a) soybean (SME) biodiesel, (b) tallow oil (TO) biodiesel, and (c) waste cooking oil (WCO) biodiesel and their blends.

acid with saturated methyl esters, the higher would be the measured viscosity values [31]. Different research studies on kinematic viscosity of biodiesel obtained the values ranging from 4.1–5.2 cSt [15,32–35] for SME

(B100), 4.7–5.5 cSt [36–40] for TO (B100), and for WCO (B100) the range is 2.7–6.6 cSt [36,41–43]. The kinematic viscosities of SME and WCO in this study were 4.912 and 5.640 cSt, respectively, which are within the range of the values obtained by the other investigators, but it is different in the case of TO (5.633 cSt). The kinematic viscosity value measured for TO was slightly higher than the published data. This could be possibly due to differences in the compositions of fatty acids in the biodiesel used by different researchers.

Biodiesel is a combination of different fatty acid methyl esters (FAMEs) such as palmitic (C16:0), stearic (C18:0), oleic (C18:1), linoleic (C18:2) and linolenic (C18:3), arachidic (C20:0), gadoleic (C20:1), behenic (C22:0), wrucic (C22:1) and lignoceric (C24:0) [15]. The weight percentage of each fatty acid varies from biodiesel to biodiesel and is dependent on the types of biodiesel source and alcohol catalysts used in the trans-esterification process. From Table 6.4, nearly 60 wt% of SME fatty acids were comprised of C18:2 and C18:3, i.e., unsaturated chains, and the rest was made up of C16:0, C18:0, and C18:1, and C20:0, C22:0, C22:1, and C24:0 contributed less than 1 wt%. However, for TO and WCO, C16:0, C18:0, and C18:1, i.e., saturated and mono-unsaturated chains, constitute for more than 75% of fatty acid mass and remaining wt% was made up of C18:2, C18:3, C20:0, C22:0, C22:1, and C24:0 [15]. Due to this wide difference in the degree of unsaturation and chain length in the FAME compositions, viscosity of SME was different from that of WCO and TO. From the viscosity values listed in Table 6.4, at 40°C, C18:0 (5.85 cSt) has high kinematic viscosity and is followed by C18:1 (4.51 cSt), C16:0 (4.38 cSt), C18:2 (3.65 cSt), and C18:3 (3.09 cSt). WCO and TCO have higher mass percentage of C20:0, C20:1, C22:0, C22:1, and C24:0 than SME, which results in high specific gravity and influences higher kinematic viscosity in the biodiesel [44,45].

TABLE 6.4

Fatty acid compositions of SME, TO, and WCO in mass percentage [25]

Fatty acids	Chemical formula	Chemical Structure	SME	TO	WCO
Lauric	$C_{12}H_{24}O_2$	C12:0	0.1	0.1	0.0
Myristic	$C_{14}H_{28}O_2$	C14:0	0.1	2.8	0.9
Palmitic	$C_{16}H_{34}O_2$	C16:0	10.2	23.3	20.4
Stearic	$C_{18}H_{38}O_2$	C18:0	3.7	19.4	4.8
Oleic	$C_{18}H_{36}O_2$	C18:1	22.8	42.4	52.9
Linoleic	$C_{18}H_{34}O_2$	C18:2	53.7	2.9	13.5
Linolenic	$C_{18}H_{32}O_2$	C18:3	8.6	0.9	0.8
Total saturated			14.1	45.6	26.1
Total unsaturated			85.1	46.2	67.2

TABLE 6.5

Low temperature kinematic viscosities (cSt) of saturated and unsaturated fatty acid compounds [28,46]

	Fatty acid methyl ester [a]									
Temperature (°C)	**C10:0**	**C12:0**	**C14:1**	**C16:0**	**C16:1**	**C18: 0**	**C18:1, Δ9c**	**C18:2**	**C18:3**	**C18:1. 12-OH**
40	1.71	2.41	2.73	4.38	3.67	5.85	4.51	3.65	3.09	15.29
35	1.87	2.69	3.04	NR*	3.96	NR*	5.08	4.08	3.32	18.58
30	2.05	2.95	3.37	NR*	4.42	NR*	5.72	4.53	3.88	23.83
25	2.23	3.29	3.71	NR*	4.94	NR*	6.44	5.03	4.07	29.77
20	2.45	3.63	4.13	NR*	5.56	NR*	7.33	5.61	4.57	37.07
15	2.71	4.07	4.73	NR*	6.38	NR*	8.51	6.43	5.14	49.47
10	3.10	4.79	5.35	NR*	7.33	NR*	9.91	7.30	5.53	64.74
5	3.49	5.45	6.13	NR*	8.55	NR*	11.66	8.47	6.59	91.92
0	4.04	nd[b]	7.01	NR*	10.15	NR*	14.03	9.84	7.33	123.83
-5	4.68	-	8.37	NR*	12.19	NR*	17.22	11.80	8.81	182.36
-10	5.40	-	9.92	NR*	14.77	NR*	21.33	14.10	10.19	271.50
LTVR	2.37	-	2.57	NR*	2.76	NR*	3.11	2.70	2.37	8.10

a Systematic (trivial) names of fatty acid methyl esters in this table in the sequence of the columns from left to right: Methyl decanoate (caprate), methyl dodecanoate (laurate), methyl 9(Z)-tetradecenoate (myristoleate), methyl 9(Z)-hexadecenoate (palmitoleate) acid, methyl 9(Z)-octadecenoate (oleate), methyl 9(Z),12(Z)-octadecadienoate (linoleate), methyl 9(Z),12(Z),15(Z) octadecatrienoate (linolenate) acid, methyl 12-hydroxy-9(Z)-octadecenoate (ricinoleate).

b Not determined. Formation of crystals at this temperature.

* NR: Not reported.

LTVR – Low Temperature Viscosity Ratio.

Therefore, the existence of higher weight percentage of C18:2 and C18:3 in SME has resulted in lower kinematic viscosity than TO and WCO.

Table 6.5, which shows the kinematic viscosity variations of various carbon chains with respect to temperature, provides a better understanding on the relation between viscosity and FAME profile. In Table 6.5, for example, if we consider the kinematic viscosity of FAME of different chain length at a temperature of 40°C, C12:0 has a kinematic viscosity of 2.41 cSt which is higher than C10:0 (1.71cSt) and kinematic viscosity of C14:1 (2.73cSt) is higher than C12:0 (2.41cSt). Therefore, it can be said that the higher the carbon chain length is, the higher kinematic viscosity would be. C18:0 (5.85 cSt) has higher kinematic viscosity value than C18:1 (4.51), C18:1 has higher value than C18:2 (3.65 cSt) and C18:3 (3.09 cSt). Consequently, saturated FAMEs (C18:0) has higher viscosity than unsaturated

FAMEs [31]. Also, with the increase in the number of double bonds in FAMEs, kinematic viscosity decreases [46]. So, SME will always have lower kinematic viscosity than WCO and TO. In this study both the pure WCO and TO and their blends with ULSD had approximately equal viscosities because of the similar fatty acid compositions [25].

6.3.3 Flash Point

Flash point is an important property of fuel for hazard management. Biodiesel flash point is suggested to be greater than the minimum limit of 130°C and 52°C for petroleum diesel, which was set by ASTM standards [12,47]. The experiments conducted on ULSD, SME, TO and WCO clearly exhibited the acceptable results that were higher than the limit. From the results displayed in Table 6.6 it is clear that TO has a high flash point of 176°C which is followed by SME and WCO with 171°C and 167°C, respectively. However, compared to biodiesels, ULSD has a lower flash point of 73±0.707°C. SME and WCO showed the similar flash point values. The flash point of biodiesel was found that it increased as the biodiesel content increased in blends. Although a high flash point value means that it is easier to handle and safer than regular diesel, it also means difficult ignition and inferior throttle response. Also, a fuel with high flash point could create carbon deposition inside engines [7].

Candeia et al. [32] measured flash point of SME biodiesel to be 168°C, which is closer to the value obtained in this current study i.e., 171°C. On

TABLE 6.6

Flash point temperatures of ULSD and its biodiesel blends

Fuel	Blend	Flash Point (°C)
ULSD	-	73.000 ± 0.707
Soybean biodiesel	B100	171.000 ± 0.000
	B50	94.000 ± 0.000
	B20	80.000 ± 0.707
	B10	79.000 ± 0.000
Tallow Oil (White grease biodiesel)	B100	176.000 ± 2.212
	B50	106.000 ± 0.000
	B20	89.000 ± 0.000
	B10	84.000 ± 0.353
Waste cooking oil biodiesel	B100	167.000 ± 0.000
	B50	96.000 ± 0.000
	B20	82.000 ± 0.000
	B10	79.000 ± 0.000

the other hand, Canakci and Sanli [7], Pereira et al. [48] and Teixeira et al. [49] measured 131°C, 150°C and 156°C, respectively. Canakci and Sanli [7] and Wyatt et al. [18] reported flash point of TO biodiesel as 150°C, while in this study it was measured 176°C. Encinar et al. [20] by experimenting with the operational parameters, such as methyl alcohol/waste cooking oil (WCO) ratio and a catalyst, created thirteen different WCO biodiesel samples, whose flash point values were measured in the range of 169–178°C. Chhetri et al. [21] reported flash point of WCO biodiesel to be 164°C. In the current report flash point was found to be 167°C, which falls within the results obtained by other researchers. The large difference between the flash point values of the biodiesel in this study and other reported values could be due to the different fatty acid profiles, which requires further studies on the fatty acid compositions of biodiesels derived from different feedstocks and different manufactures.

As shown in Figure 6.4, biodiesel content significantly influenced the flash point of the fuel. When the biodiesel content changed from 0 (ULSD) to 100% (B100), the flash point increased by average 100°C. The average slope of the flash point changes for SME, TO, and WCO biodiesels was

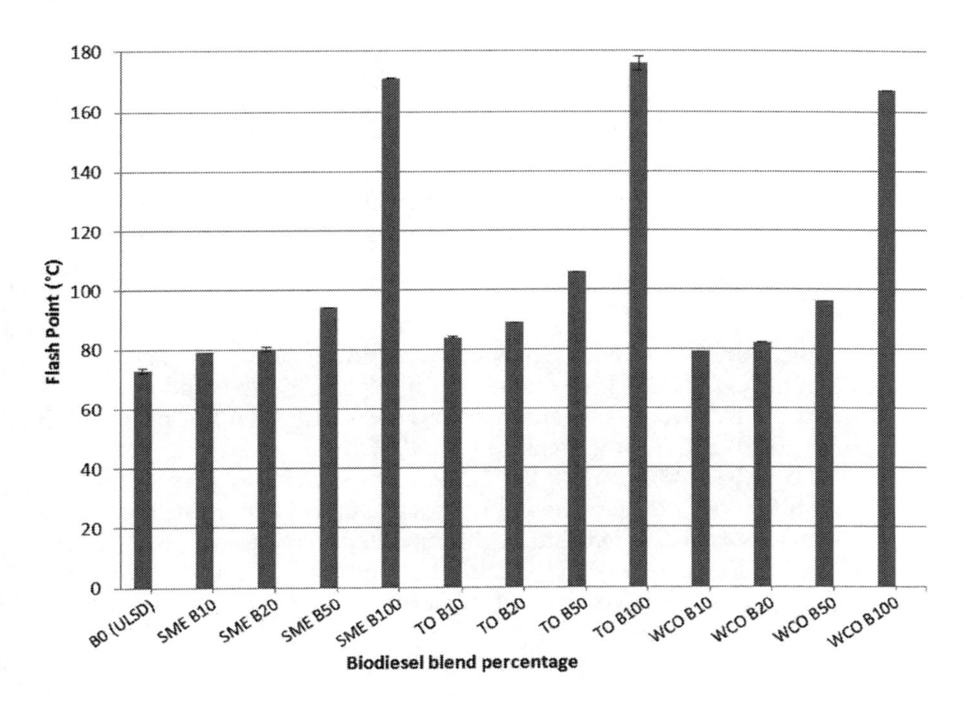

FIGURE 6.4
Flash points of ULSD and biodiesel blends.

0.964°C/BD percent. This means that a 10% increase of biodiesel in blends may increase the flash point by 10°C. The effect of biodiesel on the flash point is found to be almost an order of magnitude more significant than on the cloud point and two orders of magnitude more than on viscosity.

It is interesting to observe that the flash points of SME, TO, and WCO biodiesels changed almost the same in terms of blending, which is different from the cloud point and viscosity. It appears that the flash point of biodiesel blends is not affected by the type of feedstock. Flash point of biodiesel is known to be affected by glycerides content [21]. An increase in triglyceride content increases flash point [47], as glycerides get hydrolyzed into free fatty acids (FFA) and with the increase in FFA in the biodiesel, the flash point also increases. The presence of high molecular weight triglycerides in the biodiesel decreases the volatility of the fuel [47]. Therefore, it is assumed that the contents of glycerides in SME, TO, and WCO biodiesels were similar regardless of the type of feedstock.

The standard deviations in Tables 6.2, 6.3, and 6.6 show that the cloud point test has minimal variations in the results, and the kinematic viscosity tests and flash point tests were conducted with high precision, except a small amount of deviation, that is, ±2.212°C in TO B100 readings. Kinematic viscosities of ULSD, SME, TO, and WCO biodiesels are observed in the prescribed ranges in accordance to the ASTM D445 standards.

6.4 Conclusion

Cloud point, kinematic viscosity and flash point of three different biodiesels were measured. The cloud point and kinematic viscosity of biodiesel were higher than ULSD and increased with the increase in the percentage of biodiesel in its blend with ULSD, whereas flash points of the biodiesels were almost the same (171, 176, 167°C) within the statistical confidence interval. Viscosity of pure ULSD was found to be 3.439 cSt, and as the contents of SME, TO, and WCO increased in blends, it increased by 0.0153, 0.0226, and 0.0225 cSt/BD percent, respectively. Could point of pure ULSD was found to be -9°C, and it increased as ULSD was mixed with SME, TO, and WCO by 0.087, 0.176, and 0.117°C/BD percent. The increasing rates of flash point of biodiesels were almost the same, 10°C/BD percent regardless of the feedstock types. It is thought that the cloud point and viscosity are dependent on the contents of unsaturated fatty acids and long carbon chains. Flash point, however, is controlled by the presence of triglycerides, and the amount of triglycerides appeared to be the same for the tested biodiesels. Research data on physical properties of WCO and TO biodiesel blends were very few, and only a few data have been published on SME biodiesel blends.

Acknowledgment

This research work was funded by Mineta National Transit Research Consortium/US Department of Transportation. The authors would like to extend their sincere appreciation to Toledo Area Regional Transit Authority (TARTA), United Oil Inc., and White Mountain Biodiesel for providing us samples of ULSD and biodiesel and Dr. Hamid Omidvarborna for helping with the experiments. The views expressed in the chapter are those of the authors.

References

1. Anonyms. (2012). *Key world energy statistics-2012*. International Energy Agency.
2. Shonnard, D.R., Williams, L., Kalnes, T.N. (2010). *Camelina-derived jet fuel and diesel: Sustainable advanced biofuels*. Environ. Prog. Sustain. Energy, 29, 382–392.
3. Chiu, C.W., Suppes, G.J., Schumacher, L.G. (2004). *Impact of cold flow improvers on soybean biodiesel blend*. Biomass Bioenergy, 27, 485–491.
4. Agudelo, A., Agudelo, J., Benjumea, P. (2008). *Basic properties of palm oil biodiesel–diesel blends*. Fuel, 87, 2069–2075.
5. Kalnes, T.N., Koers, K.P., Marker, T., Shonnard, D.R. (2009). *A technoeconomic and environmental life cycle comparison of green diesel to biodiesel and syndiesel*. Environ. Prog. Sustain. Energy, 28, 111–120.
6. Zaimes, G.G., Khanna V. (2013). *Environmental sustainability of emerging algal biofuels: A comparative life cycle evaluation of algal biodiesel and renewable diesel*. Emviron. Prog. Sustain. Energy, 32, 926–936.
7. Canakci, M., Sanli, H. (2008). *Biodiesel production from various feedstocks and their effects on the fuel properties*. J. Ind. Microbiol. Biotechnol., 35, 431–441.
8. Knothe, G. (2005). *Dependence of biodiesel fuel properties on the structure of fatty acid alkyl esters*. Fuel Process. Technol., 86, 1059–1070.
9. ASTM D2500-02. (2002). *Standard test method for cloud point of petroleum products*. American Society for Testing and Materials, Pennsylvania.
10. ASTM D445-70. (1970). *Standard test method for kinematic viscosity of transparent and opaque liquids (and the calculation of dynamic viscosity)*. American Society for Testing and Materials, Philadelphia.
11. Demirbas, A. (2008). *Relationships derived from physical properties of vegetable oil and biodiesel fuels*. Fuel, 87, 1743–1748.
12. ASTM D92-05. (2005). *Standard test method for flash and fire points by Cleveland open cup tester*. American Society for Testing and Materials, Pennsylvania.
13. Liaw, H.J., Chiu, Y.Y. (2006). *A general model for predicting the flash point of miscible mixtures*. J. Hazard. Mater., 137, 38–46.
14. Faccini, C.S., Caramão, E.B., de Menezes, E.W., Krause, L.C., Da Cunha, M. E., Rodrigues, M.R.A., Moraes, M.S.A., Veses, R.C. (2008). *Tallow biodiesel: Properties evaluation and consumption tests in a diesel engine*. Energy Fuel, 22, 1949–1954.
15. Casas, A., Pérez, A., Fernández, C.M., Rodríguez, L., Ramos, M.J. (2009). *Influence of fatty acid composition of raw materials on biodiesel properties*. Bioresour. Technol., 100, 261–268.

16. Minami, E., Imahara, H., Saka, S. (2006). *Thermodynamic study on cloud point of biodiesel with its fatty acid composition*. Fuel, 85, 1666–1670.
17. Noureddini, H., Muniyappa, P.R., Brammer, S.C. (1996). *Improved conversion of plant oils and animal fats into biodiesel and co-product*. Bioresour. Technol., 56, 19–24.
18. Wyatt, V.T., Hess, M.A., Haas, M.J., Dunn, R.O., Foglia, T.A., Marmer, W.N. (2005). *Fuel properties and nitrogen oxide emission levels of biodiesel produced from animal fats*. J. Am. Oil Chem. Soc., 82, 585–591.
19. Lee, I., Johnson, L.A., Hammond, E.G. (1995). *Use of branched-chain esters to reduce the crystallization temperature of biodiesel*. J. AM. Oil Chem. Soc., 72, 1155–1160.
20. Encinar, J.M., Gonzalez, J.F., Reinares, A.R. (2005). *Biodiesel from used frying oil. Variables affecting the yields and characteristics of the biodiesel*. Ind. Eng. Chem. Res., 44, 5491–5499.
21. Chhetri, A.B., Watts, K.C., Islam, M.R. (2008). *Waste cooking oil as an alternate feedstock for biodiesel production*. Energies, 1, 03–18.
22. US NREL. (2009). *Biodiesel handling and use guidelines-2009*. US Department of Energy National Renewable Energy Laboratory.
23. Edith, O., Janius, R.B., Yunys, R. (2012). *Factors affecting the cold flow behavior of biodiesel and methods for improvement – A review*. Pertanika J. Sci. Technol., 20, 1–14.
24. Chen, G., Meng, X., Wang, Y. (2008). *Biodiesel production from waste cooking oil via alkali catalyst and its engine tests*. Fuel Proces. Technol., 89, 851–857.
25. Omidvarborna, H., Kumar, A., Kim, D.S. (2015). *NOx emissions from low-temperature combustion of biodiesel made of various feedstocks and blends*. Fuel Proces. Technol., 140, 113–118.
26. Hammond, E.G., Lee, I., Johnson, L.A., Van Gerpen, J.H., Yu, L. (1998). *The influence of trace components on the melting point of methyl soyate*. J. Am. Oil Chem. Soc., 75, 1821–1824.
27. Castillo, E., Kafarov, V., Plata, V. (2012). *Improving the low-temperature properties and filterability of biodiesel*. Chem. Eng. Trans., 29, 1243–1248.
28. Knothe, G., Steidley, K.R. (2007). *Kinematic viscosity of biodiesel components (fatty acid alkyl esters) and related compounds at low temperatures*. Fuel, 86, 2560–2567.
29. Da Cunha, M.E., Krause, L.C., Moraes, M.S.A., Faccini, C.S., Jacques, R.A., Almeida, S.R. Rodrigues, M.R.A., Caramão, E.B. (2009). *Beef tallow biodiesel produced in a pilot scale*. Fuel Process. Technol., 90, 570–575.
30. Dalai, A.K., Reaney, M.J., Bakhshi, N.N., Hertz, P.B., Lang, X. (2001). *Preparation and characterization of bio-diesels from various bio-oils*. Bioresour. Technol., 80, 53–62.
31. Allen, C.A.W., Watts, K.C., Pegg, M.J., Ackman, R.G. (1999). *Predicting the viscosity of biodiesel fuels from their fatty acid ester composition*. Fuel, 78, 1319–1326.
32. Candeia, R.A., Silva, M.C.D., Carvalho Filho, J.R., Brasilino, M.G.A., Bicudo, T. C., Santos, I.M.G., Souza, A.G. (2009). *Influence of soybean biodiesel content on basic properties of biodiesel-diesel blends*. Fuel, 88, 738–743.
33. Doll, K.M., Sharma, B.K., Suarez, P.A.Z., Erhan, S.Z. (2008). *Comparing biofuels obtained from pyrolysis, of soybean oil or soapstock with traditional soybean biodiesel: Density, kinematic viscosity, and surface tensions*. Energy Fuel, 22, 2061–2066.
34. Tate, R.E., Watts, K.C., Allen, C.A.W., Wilkie, K.I. (2006). *The viscosities of three biodiesel fuels at temperatures up to 300°C*. Fuel, 85, 1010–1015.
35. Qi, D.H., Geng, L.M., Chen, H., Bian, Y.ZH., Liu, J., Ren, X.CH. (2009). *Combustion and performance evaluation of a diesel engine fueled with biodiesel produced from soybean crude oil. Renew*. Energy, 34, 2706–2713.

36. Dias, J.M., Ferraz, C.A., Almeida, M.F. (2008). *Using mixtures of waste frying oil and pork lard to produce biodiesel*. World Acad. Sci. Eng. Technol., 2, 193–197.
37. Mata, T.M., Cardoso, N., Ornelas, M., Neves, S., Caetano, N.S. (2011). *Evaluation of two purification methods of biodiesel from beef tallow, pork lard, and chicken fat*. Energy Fuel, 25, 4756–4762.
38. Canoira, L., Gamero, M.R., Querol, E., Alcantara, R., Lapuerta, M., Oliva, F. (2008). *Biodiesel from low-grade animal fat: Production process assessment and biodiesel properties characterization*. Ind. Eng. Chem. Res., 47, 7997–8004.
39. Lebedevas, S., Vaicekauskas, A. (2006). *Use of waste fats of animal and vegetable origin for the production of biodiesel fuel: Quality, motor properties, and emissions of harmful components*. Energy Fuel, 20, 2274–2280.
40. Oner, C., Altun, S. (2009). *Biodiesel production from inedible animal tallow and an experimental investigation of its use as alternative fuel in a direction injection diesel engine*. Appl. Energy, 86, 2114–2120.
41. Phan, A.N., Phan, T.M. (2008). *Biodiesel production from waste cooking oils*. Fuel, 87, 3490–3496.
42. Felizardo, P., Correia, M.J.N., Raposo, I., Mendes, J.F., Berkemeier, R., Bordado, J.M. (2006). *Production of biodiesel from waste frying oils*. Waste Manag., 26, 487–494.
43. Meng, X., Chen, G., Wang, Y. (2008). *Biodiesel production from waste cooking oil via alkali catalyst and its engine test*. Fuel Process. Technol., 89, 851–857.
44. Atabani, A.E., Silitonga, A.S., Masjuki, H.H., Badruddin, I.A., Mekhilef, S., Mahlia, T.M.I. (2012). *A comprehensive review on biodiesel as an alternative energy resource and its characteristics*. Renew. Sust. Energ. Rev., 16, 2070–2093.
45. Yuan, W., Hansen, A.C., Zhang, Q. (2008). *Predicting the temperature dependent viscosity of biodiesel fuels*. Fuel, 88, 1120–1126.
46. Knothe, G. and Steidley, K.R. (2005). *Kinematic viscosity of biodiesel fuel components and related compounds. Influence of compound structure and comparison to petrodiesel fuel components*. Fuel, 84, 1059–1065.
47. Fernando, S., Karra, P., Hernandez, R., Jha, S.K. (2007). *Effect of incompletely converted soybean oil on biodiesel quality*. Energy, 32, 844–851.
48. Pereira, R.G., Oliveira, C.D., Oliveira, J.L., Oliveira, P.C.P., Fellows, C.E., Piambam, O.E. (2007). *Exhaust emissions and electric energy generation in a stationary engine using blends of diesel and soybean biodiesel*. Renew. Energy, 32, 2453–2460.
49. Teixeira, L.S.G., Couto, M.B., Souza, G.S., Filho, M.A., Assis, J.C.R., Guimaraes, P.R.B., Pontes, L.A.M., Almeida, S.Q., Teixeira, J.S.R. (2010). *Characterization of beef tallow biodiesel and their mixtures with soybean biodiesel and mineral diesel fuel*. Biomass Bioenergy, 34, 438–441.

7

Life Cycle Assessment of Biodiesel

Qingshi Tu

Acronyms

Acronym	Definition
AEZ-EF	Agro-Ecological Zone Emission Factor
ALCA	Attributional life cycle assessment
ANL	Argonne National Laboratory
AusLCI	Australian National Life Cycle Inventory Database
CCLUB	Carbon Calculator for Land Use Change from Biofuels Production
CARD	Center for Agricultural and Rural Development at Iowa State University
CLCA	Consequential life cycle assessment
CLCD	Chinese Life Cycle Database
EIO-LCA	Economic input-output (EIO)-based life cycle assessment
ELCD	European reference Life Cycle Database
EROI	Energy return on investment
FAME	Fatty acid methyl esters
FAPRI	Food and Agricultural Policy and Research Institute
FASOM	Forestry and Agricultural Sector Optimization Model
GHG	Greenhouse Gas
GNOC	Global crop- and site-specific nitrous oxide emission calculator
GREET	Greenhouse Gases, Regulated Emissions, and Energy Use in Transportation model
GTAP	Global Trade Analysis Project
GWP	Global warming potential
IPCC	Intergovernmental panel on climate change
LCFS	Low Carbon Fuel Standard
LUC	Land use change
iLUC	Indirect land use change
dLUC	Direct land use change
MOVES	Motor Vehicle Emission Simulator
NREL	National Renewable Energy Laboratory

OPGEE	Oil Production Greenhouse gas Emissions Estimator
RED	Renewable Energy Directive
PEF	Product Environmental Footprint
RFS	Renewable Fuel Standard
USDA	United States Department of Agriculture
US EPA	United States Environmental Protection Agency
USEEIO	United States Environmentally Extended Input-Output Model

7.1 Introduction

Life cycle assessment (LCA) is a methodology that enables the evaluation of both direct and indirect environmental impacts of a product throughout its production, use and end-of-life (EoL) management (Life cycle assessment – Requirements and guidelines 2006). An LCA starts with a clear definition of goal (e.g., comparing the GHG emissions, on an energy-basis, between biodiesel produced from soybean and petroleum-based diesel in the US) and scope (e.g., system boundary) of the study (Figure 7.1). Inventory analysis quantifies the flows of materials, energies, products, waste and emissions throughout the life cycle of the product. Life cycle impact assessment (LCIA) quantifies the potential environmental impacts from the life cycle inventory (LCI) data. The environmental impacts are grouped into several categories, based on the stage at which the cause-effect relationship between inventory and environment impacts is evaluated (i.e., midpoint or endpoint). Midpoint impact assessment focus on individual environmental problems such as climate change and acidification. Other common midpoint

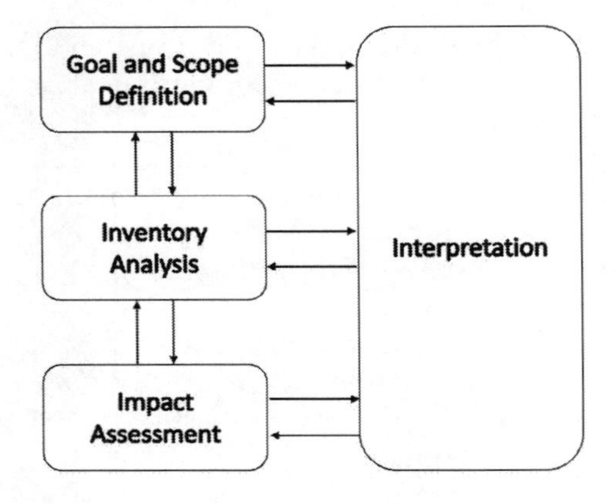

FIGURE 7.1
Life cycle assessment framework from ISO 14044–2006 (Life cycle assessment – Requirements and guidelines 2006).

impact categories include terrestrial ecotoxicity, ozone depletion, freshwater eutrophication, and so on. Endpoint impact assessment characterizes the midpoint impacts further into potential damage to human health, ecosystems and resources. The exact number of impact categories varies among different impact assessment methods. A summary of major midpoint and endpoint impact assessment methods and impact categories included in each method can be found in (Acero et al. 2015). Interpretation of LCA results enables the understanding of environment impacts of the product, as well as the comparison with alternatives (Curran 2016).

For most crop-based biofuels (e.g., biodiesel produced from soybean), the life cycle typically starts with resource extraction activities that lead to the production of multiple inputs to the feedstock growth, including mineral fertilizers, electricity from fossil energy carriers, land management for crop growth and water for irrigation. The biomass is harvested and processed for production of biofuel and co-products. The biofuel is distributed through fuel stations and used in on-/off-road engines. Detailed modeling steps and LCI data of each life cycle stage (square boxes in Figure 7.2) are dependent upon the goal and scope of the biofuel LCA study.

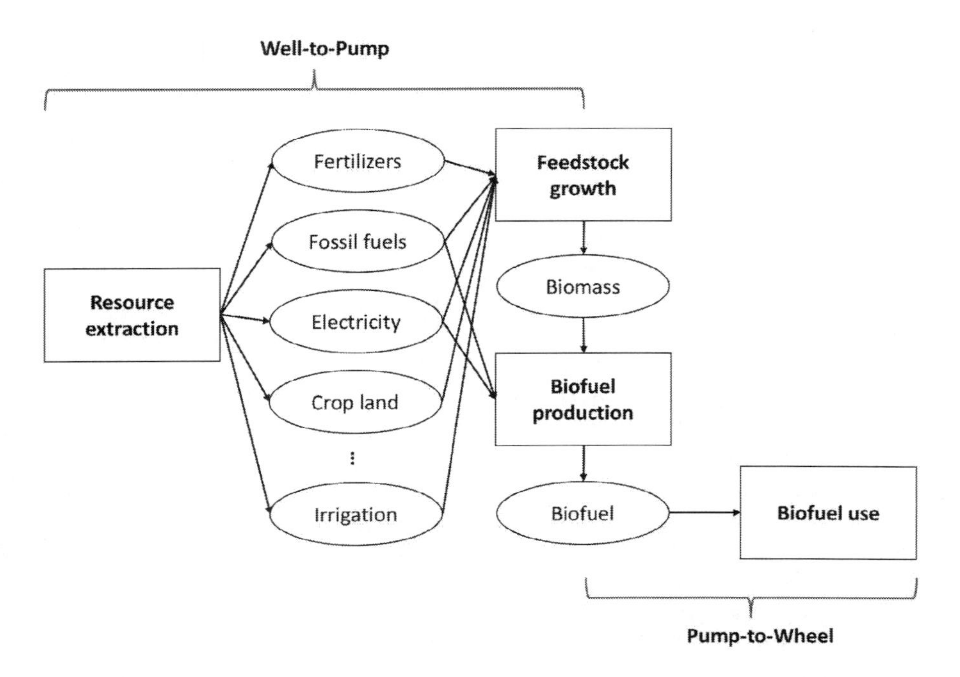

FIGURE 7.2
Illustration of system boundary of crop-based biofuels.

7.2 Goal and Scope of Biodiesel LCAs

7.2.1 Functional Unit, System Boundary, and Impact Categories

Functional unit (FU) defines the function of the system and provides a reference for quantifying the inputs and outputs related to it. Another purpose of a FU is to enable comparison between different systems. Typically, biodiesel LCAs use volume-based (e.g., 1 liter), mass-based (e.g., 1 kg), or energy content-based (e.g., 1 MJ) FUs. The environmental impacts of different fuels (e.g., gasoline *vs* biodiesel) can be compared based on these FUs directly. However, since these FUs focus on fuel properties alone (i.e., well-to-pump), adjustment for factors affecting the use of fuels (e.g., difference in fuel economy between gasoline- and diesel-powered vehicles) may be needed when these FUs are used in LCAs conducted for policy design (Farrell 2007).

System boundary decides which processes to be included in the analysis throughout the life cycle. Certain upstream processes (e.g., coal mining) are usually included, as they are the inputs to the inventories that are ubiquitously used throughout the life cycle (e.g., electricity). Others may only be included if a specific type of feedstock is used for biodiesel production. For example, upstream processes such as mineral extraction for fertilizer production are typically included for crop-based feedstocks (e.g., soybean), while are excluded by waste feedstocks such as animal fats in most cases. The feedstock production stage for crop-based feedstocks mainly involve land management (e.g., tillage), fertilizer use, irrigation and harvesting. For waste feedstocks, pretreatment processes, such as oil separation from grease trap waste (Tu and McDonnell 2016), are typically considered. The environmental burdens from transportation of materials (e.g., fertilizers, chemicals), feedstocks, intermediate (e.g., soy oil from extraction) and final biodiesel products are usually very small and neglected by many studies. Another common exclusion is the construction, maintenance and EoL management of infrastructure, with the assumption that the normalized environmental impacts (e.g., for 1 MJ of biodiesel) are usually negligible due to a long lifetime of infrastructure, although the limitation of such an assumption has been indicated (Canter et al. 2014). Currently, alkali-catalyzed transesterification is the major technology for the biodiesel industry, and most LCA studies adopt it as a default option for modeling biodiesel production. Innovations have been proposed on transesterification technologies, including heterogeneous catalysis, *in-situ* transesterification, ultrasound-assistance and supercritical conditions. LCA studies have been performed for some of these technologies to compare their environmental impacts with alkali-catalyzed transesterification (Kiwjaroun et al. 2009; Morais et al. 2010; Kiss et al. 2014; Tuntiwiwattanapun et al. 2017).

A majority of biodiesel LCA studies focus on the midpoint impact assessment (Collet et al. 2015), a large number of which focus on climate change using the global warming potential (GWP) factors of a 100-year time frame (i.e., GWP_{100}; hereafter GWP for simplicity). GWP of a GHG is expressed in the form of CO_2-equivalent value which is calculated based on its capability to absorb infrared radiation and its life time in the atmosphere (US EPA 2016). Water use and water quality are another major area of interest (Subhadra and Edwards 2011; Yang et al. 2011; Tu et al. 2016), while other impact categories are relatively less investigated.

7.2.2 Land Use Change

Land use change (LUC) is an important factor to consider when modeling life cycle environmental impacts of biodiesel. For example, LUC may cause both direct (e.g., microbial nitrification and denitrification of fertilizers) and indirect (e.g., volatilization, leaching and runoff) GHG emissions from agriculture activities, leading to a significant amount of N_2O release. The change of land use due to an increased demand of biodiesel could occur in both direct and indirect ways. Direct land use change (dLUC or sometimes simply denoted by LUC) refers to the situations where, for example, a land that previously was not used for the growth of biodiesel feedstock (e.g., soybean) is now used to grow the feedstock, as a response to the increased market price of soybean. Indirect land use change (iLUC) refers to the situations where the change of land use is a result of "ripple effect" from the change in demand of biofuels. For example, increased demand of corn for ethanol (due to biofuel mandate) leads to the increased production of corn and reduction in soybean production in the US, which in turn reduces the export of soybean from the US to Brazil, leading to conversion of pasture lands to grow soybean in Brazil. This land use change results in the reduced production of beef and hence beef prices increase, which could cause deforestation for more pasture lands, leading to a significant release of carbon from soil and biomass (Rajagopal and Zilberman 2010).

Many biodiesel LCA studies exclude LUC from the scope of study, due to lack of modeling capacity and/data to accurately model the drivers and causality of LUC, particularly those of iLUC. On the other hand, biofuel policies often strive to include the impacts from potential LUC in order to reduce the unintended consequences, despite those significant uncertainties of modeling results still exist (see Section 7.5 for more details).

7.2.3 Modeling Approaches

There two major modeling approaches for LCA studies in general: process-based and economic input-output (EIO)-based approaches (Hendrickson et al. 1997). Process-based approach is relatively easy to interpret, could have a high level of details regarding individual processes and hence is widely used for biodiesel LCAs. The flows of materials, energy, products and waste can be explicitly tracked through the system boundary and accordingly, the major contributors to the total environmental impacts can be identified. One major disadvantage of process-based approach is the introduction of truncation error (i.e., the proportion of environmental impacts that are not accounted for) due to a limited system boundary. Although truncation error could be managed by setting a small cut-off threshold for individual processes (e.g., a process that contributes less than 0.1% of total environmental impacts is excluded from system boundary), the cumulative truncation error of all relevant processes could still be considerable (Ward et al. 2018).

EIO-based LCA (EIO-LCA) quantifies the life cycle inventory data in monetary terms using an IO table that contains monetary transactions between economic sectors of a country (***A*** matrix in Equation 7–1). The total output (i.e., life cycle inventory) from all economic sectors (***x*** vector) in response to the total demand of products (***y*** vector) is calculated using Leontief inverse ($[\boldsymbol{I}\text{-}\boldsymbol{A}]^{-1}$). By multiplying the total output with emission factors (*f* row vector) for respective economic sectors, the total emissions (*e*) can be calculated for a given total demand of products.

$$e = fx = f[I - A]^{-1}y \tag{7–1}$$

EIO-LCA captures the environmental impacts from nonphysical activities associated with biodiesel production, such as investment and insurance for biodiesel plants, that are typically not included in the system boundary of process-based LCAs and therefore, represents a more complete system boundary. Another advantage of EIO-LCA is the capture of "circular" nature of the life cycle through the Leontief inverse. For example, the production of electricity for transesterification process will require a certain amount of electricity itself (e.g., for certain operations of the power plant). These two advantages combined could reduce the truncation error compared to process-based LCA approach. On the other hand, limitations of the EIO-LCA approach include: 1) low level of details on individual processes, as the transaction data is aggregated at sector-level (e.g., power generation and supply sector represents electricity supplied from both carbon-intensive technologies such as coal-fired plants and low-GHG technologies such as hydropower) and the corresponding emission results represent the average of all technologies involved; 2) the inter-sectoral monetary coefficients in the IO table are fixed and often dated. For example, the transaction data for constructing the monetary coefficients in the IO table of US economy is updated every 4–5 years by US Bureau of Economic Analysis, therefore the recent change of technology (e.g., improved yield for oil extraction, less energy-intensive separation process) may not be accounted for using EIO-LCA; 3) the development of a complete set of sector-level emission factors remains a challenge, given both strenuous efforts needed for data collection and uncertainties from the heterogeneity of technologies involved in a sector (Yang et al. 2017b); and 4) emissions from use of products are not included. ECO-LCA is often used for screening-level or aggregated (economy)-level analyses.

A third modeling approach, taking advantage of the high specificity of process-based LCA and the more complete system boundary of EIO-LCA, is called hybrid LCA. The process-based data is in physical units and represented in a technology matrix while the IO data is in monetary unit. These two sets of data are linked through supply-use tables (SUTs) in the hybrid LCA (Bilec et al. 2006). Hybrid LCA has been applied to biodiesel and the results indicate that truncation error could be considerably reduced (Acquaye et al. 2011), although the uncertainties risen from aggregated data of IO sectors may offset the accuracy gained from using process-based data in some cases (Yang et al. 2017a).

7.3 Tools and Data for Biodiesel LCAs

7.3.1 LCA-Based Tools

The Greenhouse Gases, Regulated Emissions, and Energy Use in Transportation (GREET) Model is developed by Argonne National Laboratory (ANL) in the US and is one of the most widely used analytical tools for GHG emissions from transportation. GREET consists two sets of submodels, namely GREET1 and GREET2 series, that are specialized in modeling fuel cycle ("well-to-pump") and vehicle cycle ("pump-to-wheel" plus vehicle production and recycling), respectively. GREET1 contains over 100 fuel pathways among which biodiesel from transesterification includes those from soybean, palm, rapeseed, jatropha, camelina, used cooking oil, tallow, algae, and so on. In addition to GHG emissions from production of both biodiesel (e.g., transesterification) and associated material and energy inputs, GREET also uses "Carbon Calculator for Land Use Change from Biofuels Production" (CCLUB) to estimate GHG emissions from LUC induced by change in demand of biofuels. The current version of CCLUB contains four LUC scenarios based on differences in soybean biodiesel demand and modeling assumptions.

GHGenius is developed and maintained by S&T consultants Inc. in Canada (www.ghgenius.ca). GHGenius is based on a model originally created by Dr. Mark Delucchi at University of California, Davis, in the 1980s for estimating transportation emissions. GHGenius is an Excel spreadsheet-based tool that evaluates the life cycle GHG emissions from production and use of fossil and alternative fuels. The tool contains over 200 fuel pathways, many of which are represented by Canada-specific data. GHGenius can perform LCAs using region-specific (e.g., east, central or west) data for Canada, US and EU, as well as country-level data for Mexico, India and New Zealand. GHGenius applies IPCC Tier 2 emission factors for modeling dLUC (Bonomi et al. 2017).

BioGrace GHG calculation tool (BioGrace I) allows the users to reproduce the default GHG emission values for biofuel pathways that are specified in Annex V of the EU Renewable Energy Directive (RED) (www.biograce.net). BioGrace I is also an Excel spreadsheet-based tool and the built-in pathways for biodiesel include soybean, rapeseed, sunflower, palm and waste vegetable oil/animal fat. For each pathway, the users are also able to update the default values in each life cycle stage (e.g., soybean yield, electricity use for oil extraction) to adapt the calculation for a specific scenario of interest. BioGrace I applies IPCC Tier 1 emission factors for modeling dLUC (Bonomi et al. 2017).

7.3.2 Data Sources and Harmonization

The data for an LCA can be categorized into primary and secondary sources. Primary data is often collected directly from processes (e.g., transesterification), facilities (e.g., biodiesel plants), companies, and so on. The collection of primary data can be time-consuming and cost-prohibitive in some cases and therefore, secondary data which is compiled in publications and databases, is often used in LCA studies in lieu of primary data. Data could also be categorized into foreground and background data. The former refers to the data that is collected specifically for the study of interest, such as the electricity consumption data for a particular oil extraction technology. The latter refers to the generic data that does not bear the specific context of the study (e.g., material and energy inputs for electricity generation which is only dependent upon technology mix of a given region).

There are several major databases used for LCA in general. Ecoinvent database (www.ecoinvent.org) provides documented process data for thousands of products, and is one of the most widely used databases for LCA. The current version (v3.5) of ecoinvent contains process inventory data for soybean biodiesel production with the geographical scopes of the US, Brazil, and the global average. Another comprehensive database is GaBi that is developed by thinkstep (www.thinkstep.com), which contains over 12,000 LCI datasets. GaBi contains the process inventory data for biodiesel produced from rapeseed in EU. US Life Cycle Inventory (USLCI) is a US-specific LCI database maintained by NREL and its partners (www.nrel.gov/lci/). European reference Life Cycle Database (ELCD) is established by EU Commission's Joint Research Centre (JRC) with over 500 datasets for industries such as chemicals, metals and energy (http://eplca.jrc.ec.europa.eu/ELCD3/). Chinese Life Cycle Database (CLCD), developed by Professor Hongtao Wang's group at Sichuan University, China, is a leading database with over 600 datasets that represent average technologies for multiple sectors in China, including energy, transportation, metal, chemicals, and so on. Other similarly country-/region-specific LCI databases include Australian National Life Cycle Inventory Database (AusLCI) and Quebec LCI database (Lesage and Samson 2016). These comprehensive databases provide a good coverage of background data (e.g., electricity generation, material production and transportation) for biodiesel LCAs. For EIO-LCA, single-region IO databases such as the USEEIO (Yang et al. 2017b) and multi-region IO databases such as EXIOBASE (Wood et al. 2015) provide different levels of sectoral and spatial coverages.

There are also country-specific databases that contain many detailed data on agricultural activities, which can potentially be used for modeling biodiesel feedstock production with a high spatial resolution. The LCA Commons database (www.lcacommons.gov/) is developed and maintained by the USDA National Agricultural Library (NAL), which complies US-specific agricultural process data (e.g., operation of a 325-horsepower soybean header over 10,000 m^2 in Ohio, US). Other similar databases include AGRIBALYSE (France-context) and Agri-footprint (EU-context) (Corrado et al. 2018).

In some cases, rather than data gaps, LCA practitioners may face the issue of data variations, which could also lead to considerable uncertainties in the results. Data variations could be a reflection of the inherent uncertainty (e.g., range of moisture levels in dewatered algae from a specific centrifuge due to limitation of its design) or an artifact of data preparation (e.g., arbitrary assumptions of CH_4 generation from anaerobic digestion (AD) of defatted algae without considering its composition), yet in most cases, a mix of both. A complete separation of the two sources is unlikely, nonetheless, harmonization has been performed for many renewable energy (Dolan and Heath 2012; Kim et al. 2012; Asdrubali et al. 2015) and biofuel LCA studies (Farrell et al. 2006; Davis et al. 2012; Tu et al. 2018) in order to reach a more consistent conclusion of their benefits. In order to facilitate the meta-analysis and harmonization of data from different sources, it is important to document the data preparation in a transparent way and ensure the accessibility of raw data from which final data is prepared (Hertwich et al. 2018). Data variations could be reduced through several harmonization steps, including: (1) disaggregation of LCI into unit process level. LCI data from different sources may be presented in different levels of aggregation. For example, energy consumption for algae growth may be in the unit of MJ/kg algal oil, which is affected by the "hidden" assumptions regarding the yields of algae harvesting, dewatering and oil extraction steps. Rather, by disaggregating such data into unit process level for each step, that is, MJ/kg algae before harvesting, MJ/kg algae harvested, and MJ/algal oil, it is possible to harmonize the assumptions regarding energy consumption specific to algae growth alone; (2) unifying the calculation methods for intermediate results. For example, the areal yield of algae during growth stage could be calculated through sophisticated growth model (Leow et al. 2018) or using empirical values (e.g., 25 $g/m^2/d$). In this case, applying a consistent calculation method for different data sources regarding algae yield may considerably reduce the variation in LCI results; (3) filling missing steps. Some data sources may neglect certain important steps, e.g., energy consumption for treating flue gas to reach desired CO_2 concentration for algae growth, therefore bridging such gaps using consistent data may reduce the variation between data sources; (4) conserving the dependency between steps. For instance, the CH_4 generation rate from AD of defatted algae should be determined based on its composition of protein and carbohydrates which is related to algae species and growth condition (Tu et al. 2017).

7.4 Climate Change Impacts of Biodiesels by Feedstock

7.4.1 Soybean and Rapeseed

Soybean is a major biodiesel feedstock for many countries, including the US, Argentina, and Brazil. The life cycle of soybean biodiesel is a typical example of the biofuel life cycle shown in Figure 7.2. There has been a long history of LCA research on soybean biodiesel and the reported GWP results (e.g., 23–137 g CO_2-eq/MJ (Castanheira et al. 2015; Cerri et al. 2017; Chen et al. 2018)) are typically lower than that of petroleum-based diesel (e.g., 94 g CO_2-eq/MJ (Chen et al. 2018)). The variations in reported results are mainly caused by technological differences (e.g., difference in carbon intensity of electricity or soil management practices between US and Brazil), as well as the uncertainties in modeling potential LUC and associated GHG emissions from increased demand of soybean biodiesel (see Section 7.5 for more details). The energy return on investment (EROI) of soybean biodiesel ranges from 1.8 to 5.5 MJ/MJ (Sheehan et al. 1998; López et al. 2010; Pradhan et al. 2011; Piastrellini et al. 2017), higher than that of petroleum-based diesel (e.g., 0.83 MJ/MJ (Sheehan et al. 1998)).

Rapeseed is a major biodiesel feedstock in Europe. Similar to soybean biodiesel, the GWP results of rapeseed biodiesel also have a wide range, for example, 15–170 g CO_2-eq/MJ (Malça and Freire 2011), and the EROIs range from 1.7 to 2.6 (Firrisa et al. 2014; van Duren et al. 2015).

7.4.2 Palm Oil

Palm oil is a major biodiesel feedstock for countries in Southeast Asia, such as Indonesia, Malaysia and Thailand. Some LCA studies suggested that palm biodiesel possessed the advantages of high EROI (e.g., 3.2–7.8 MJ/MJ (Yee et al. 2009; Harsono et al. 2012)) and low GHG emissions (e.g., 4–126 g CO_2-eq/MJ (Thamsiriroj and Murphy 2009; Hansen et al. 2014; Castanheira and Freire 2017)). However, other studies have found that increasing the acreage of palm cultivation for biodiesel production could lead to significant LUC and deforestation, which could cause substantial GHG emissions and loss of biodiversity (Fitzherbert et al. 2008; Vijay et al. 2016). For this reason, European Parliament has voted to phase out the use of palm oil for transportation fuel from 2030 (Reuters 2018) and initiatives have been proposed for sustainable palm oil production (United Nations Development Programme 2012; Tropical Forest Alliance 2017).

7.4.3 Jatropha Curcas

Jatropha curcas ("jatropha" hereafter for simplicity) is considered an alternative to crop-based feedstocks, given its low requirement for fertilization and irrigation. Jatropha has been an emerging feedstock in India. The life cycle of jatropha biodiesel is similar to those of crop-based feedstocks. Despite the potential to grow on marginal lands with minimal management, the yield of jatropha fruit could be negatively affected by such a low-input cultivation practice (Achten et al. 2007; Portugal-Pereira et al. 2016). The different cultivation practices (e.g., rainfed vs. irrigated), in conjunction with different assumptions of how stem, hull and husk of jatropha fruit are used, have led to mixed conclusions on impacts of jatropha biodiesel. For instance, the EROI results range from 1.2 to 8.6 MJ/MJ (Kumar et al. 2012) and range of GWPs is between -93 and 194 g CO_2-eq/MJ (Achten et al. 2010; Bailis and Baka 2010; Portugal-Pereira et al. 2016).

7.4.4 Waste Fats, Oils and Greases (FOGs)

Used cooking oil (UCO) accounts for a large share of feedstock supply in the US, and is the main biodiesel feedstock for other countries such as China, Japan, and New Zealand (Kim et al. 2018). UCO often contains a free fatty acid (FFA) level around 5% or less, which is commonly treated with acid-catalyzed esterification to convert FFA into corresponding fatty acid methyl esters (FAME) (Chai et al. 2014). Although UCO biodiesel has a mature production pathway, the results of its LCA still vary considerably. For example, the GWP results of UCO biodiesel range from 6.2 to 75 g CO_2-eq/MJ (Quek and Balasubramanian 2014; Caldeira et al. 2016), mainly due to different assumptions regarding energy consumption during UCO pretreatment and different emission factors for energy supplies.

Waste animal fats, including tallow, lard and poultry fat, are by-products from meat processing facilities. Waste animal fats typically have a higher FFA concentration than UCO and a fatty acid profile with high concentrations of saturated fats (Adewale et al. 2015). The biodiesel production pathway for waste animal fats is similar to that of UCO, except that the material and energy inputs for rendering step are always included. There are few

LCA studies on biodiesel from waste animal fats. For tallow biodiesel, the EROI results and GWP are 2.2–5.7 MJ/MJ (Nelson and Schrock 2006; López et al. 2010) and 23–54 g CO_2-eq/MJ biodiesel (Dufour and Iribarren 2012; Quek and Balasubramanian 2014), respectively. For poultry fat, the EROI results and GWP are 1.94 MJ/MJ (López et al. 2010) and 23 g CO_2-eq/MJ biodiesel (Dufour and Iribarren 2012), respectively. Dependent upon the scope definition, a proportion of environmental impacts from animal production (e.g., raising a cattle) may also be included as a part of upstream impacts (Nelson and Schrock 2006; López et al. 2010). Such a change in system boundary, however, could lead to a significant reduction in the advantages of waste animal fats. For example, the EROI could drop to 0.2–0.9 MJ/MJ when part of the impact from cattle growth is attributed to biodiesel production (Nelson and Schrock 2006; López et al. 2010).

Other sources of waste FOGs, such as grease trap waste, are not utilized at a large scale due to their heterogeneous properties. The FOG content of such feedstocks is often not stable and the FOG is usually contaminated with fine solids and emulsion (e.g., from grease trap cleaning), causing a significant difficulty in extracting high-purity FOG for biodiesel production. Several technologies have been proposed to improve the utilization of FOG from grease trap waste, including pretreatment by hydrolysis (Montefrio et al. 2010) and extraction by waste cooking oil (Tu et al. 2016). Existing LCA studies show that biodiesel produced from grease trap waste could have lower life cycle GHG emissions compared to those from soybean, although achieving such reductions would require a certain level of feedstock quality and treatment (e.g., high FOG yield, low FFA concentration) and a proper use of byproducts (e.g., heat and electricity generation from burning biogas produced from anaerobic digestion of residual waste) (Hums et al. 2016; Tu and McDonnell 2016).

7.4.5 Algae

Algae, particularly microalgae, have been an emerging feedstock for biodiesel, given its potentially high oil productivity. Algae growth does not compete with other crops for arable lands and can potentially reduce the freshwater use, compared to crop-based feedstocks, by utilizing saline or wastewater. Yet, the large-scale commercialization of algae biodiesel is still facing several challenges, including maintaining monoculture growth environment, preventing pond crash (Sandia National Laboratory 2013), high energy consumption for algae dewatering (Shuba and Kifle 2018) and effective separation of valuable co-products (Kwan et al. 2016).

A wide variety of technologies and relevant assumptions have been investigated in the literature of algae biodiesel, given the still evolving R&Ds and their nascent status of commercial application. Such a diversity in technological assumptions (e.g., different choices of lipid extraction technologies and assumptions of corresponding energy consumptions), in conjunction with variations in methodological choices of LCA conducted by individual researcher(s), have led to significant uncertainties regarding the energy and climate change impacts of algae biodiesel (Handler et al. 2012; Menten et al. 2013; Sills et al. 2013; Collet et al. 2015). For example, although the majority of the LCAs assume flue gas as the source of CO_2 supply for photosynthesis of algae, the burdens related to flue gas treatment (e.g., increasing CO_2 concentration by chemical sorption) and delivery are often not accounted for in the system boundary (Tu et al. 2017). In addition, algae productivity and lipid concentration are dependent upon the choice of growth modeling, which could lead to a significant variation in final results between studies.

Efforts have been made to reduce the uncertainty of LCA results for algae biodiesel, by identifying and harmonizing key factors related to scope definition, methodological and technological choices and data sources. Liu et al. (2012) harmonized functional unit, system boundary and choice of co-product offset for six studies, which reduced the range of results (among these studies) for life cycle energy consumptions by 83% and a similar reduction (76%) was achieved for GHG emissions. The harmonized results are 1.4 MJ/MJ algae biodiesel and 0.19 kg CO_2-eq/km traveled by algae biodiesel, respectively. Tu et al. (2018), on the other hand, focused on the harmonization of data and assumptions of major technological options throughout the algae biodiesel life cycle. The authors proposed a baseline combination of technological options that are considered most common in literature, including a raceway system, direct injection of flue gas, N&P supply by synthetic fertilizers, average US electricity grid and climate condition, flocculation for algae harvesting, centrifuge for algae dewatering, wet hexane extraction method for lipid extraction, alkali-catalyzed transesterification, and anaerobic digestion for defatted algae. By applying the harmonized data and assumptions for these technological options, the uncertainty range of model results were significantly reduced through minimizing the artifact of data preparation from different sources. The inter-quartile ranges (IQR) of GWP and fossil energy consumption are reduced by 46% and 63%, respectively, as compared to the results reported in literature. The corresponding median values are 99 g CO_2-eq/MJ and 3.5 MJ/MJ, respectively.

Figure 7.3 summarizes the EROI and GWP values for different feedstocks from selected studies. The average EROI of all studies is 3.7 MJ/MJ and most of values are larger than 1 MJ/MJ, indicating an energy gain as compared to petroleum-based diesel (e.g., 0.83 MJ/MJ (Sheehan et al. 1998)). The average GWP is 80 g CO_2-eq/MJ and most of the values are below that of petroleum-based diesel (e.g., 94 g CO_2-eq/MJ (Chen et al. 2018)). For crop-based feedstocks, difference in LUC modeling assumptions contributes a significant share to the uncertainties in GWP results. For waste FOGs, the heterogeneous properties of grease trap waste could lead to significant uncertainties in estimating GWP of its biodiesel. Given a of wide selection of technologies and variations in inventory data, algae biodiesel is characterized by a large uncertainty in both EROI and GWP. Therefore, to reduce the uncertainties, it is

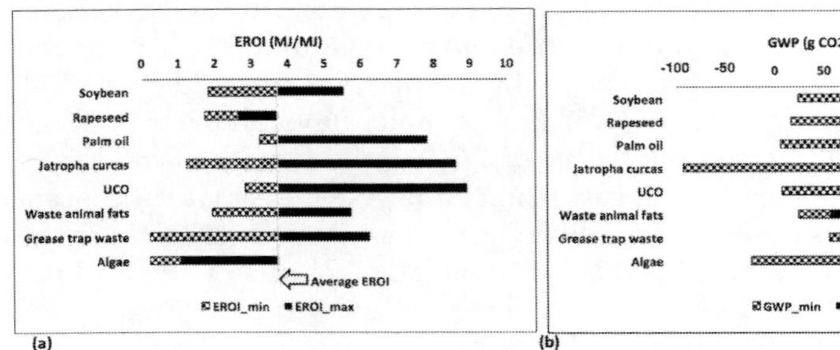

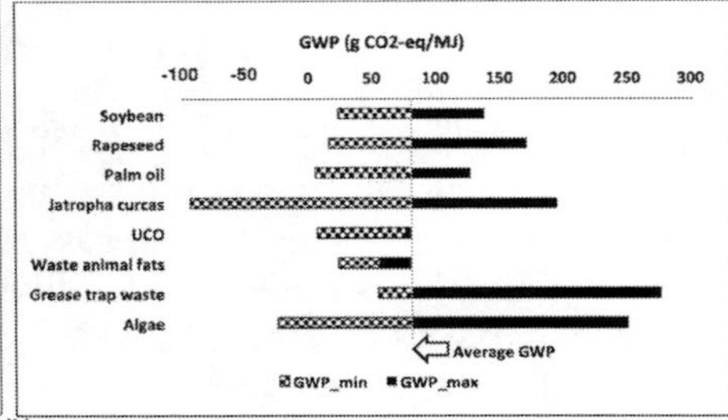

FIGURE 7.3
Illustration of uncertainties in EROI (4a) and GWP (4b) results for biodiesel produced from different feedstocks in the US.

*Note 1: the results are based on selected LCA studies, which is only for illustrative purpose.
**Note 2: results are converted into a common energy basis (i.e., per MJ biodiesel), based on density and LHV of biodiesel being 0.88 kg/L and 38.5 MJ/kg (Mehta and Anand 2009), respectively; no harmonization is performed on methodological choices (e.g., applying a consistent allocation method for all studies) or technological choices (e.g., applying the same assumption for the use of stem, hull and husk of jatropha fruit for all jatropha biodiesel studies).
***Note 3: Average EROI/GWP = average value from all data used for this figure.
****Sources of data used for the figure:
-Soybean: EROI_low = (Piastrellini et al. 2017), EROI_high = (Pradhan et al. 2011), GWP_low = (Cerri et al. 2017), GWP_high = (Castanheira et al. 2015)
-Rapeseed: EROI_low = (Firrisa et al. 2014), EROI_high = (Firrisa et al. 2014), GWP_low = (Malça and Freire 2011), GWP_high = (Malça and Freire 2011)
-Palm oil: EROI_low = (Harsono et al. 2012), EROI_high = (Yee et al. 2009), GWP_low = (Castanheira et al. 2015), GWP_high = (Hansen et al. 2014)
-Jatropha curcas: EROI_low = (Kumar et al. 2012), EROI_high = (Kumar et al. 2012), GWP_low = (Portugal-Pereira et al. 2016), GWP_high = (Achten et al. 2010)
-UCO: EROI_low = (López et al. 2010), EROI_high = (Talens Peiró et al. 2010), GWP_low = (Caldeira et al. 2016), GWP_high = (Quek and Balasubramanian 2014)
-Waste animal fats: EROI_low = (López et al. 2010), EROI_high = (Nelson and Schrock 2006), GWP_low = (Dufour and Iribarren 2012), GWP_high = (Quek and Balasubramanian 2014)
-Grease trap waste: EROI&GWP = (Tu and McDonnell 2016)
-Algae: EROI&GWP = (Tu et al. 2018).

crucial to harmonize both technological choices of production pathways and modeling choices of LUC.

7.5 Application of LCA in Biofuel Policies

7.5.1 Low Carbon Fuel Standard (LCFS)

Low Carbon Fuel Standard (LCFS) regulates the carbon intensity (CI) of transportation fuels that are used in California and the goal is to reduce the CI of transportation fuel pool by at least 10% compared to a 2010

TABLE 7.1

LCFS compliance schedule for 2010–2020

Year	Average carbon intensity (CI) for gasoline and fuels used as substitute for gasoline (gCO_2e/MJ)	Average carbon intensity (CI) for diesel and fuels used as substitute for diesel (gCO_2e/MJ)
2010	Reporting only	Reporting only
2011	95.61	94.47
2012	95.37	94.24
2013	97.96	97.05
2014	97.96	97.05
2015	97.96	97.05
2016	96.50	99.97
2017	95.02	98.44
2018	93.55	96.91
2019	91.08	94.36
2020 and subsequent years	88.62	91.81

Source: www.arb.ca.gov/regact/2015/lcfs2015/lcfsfinalregorder.pdf

baseline by 2020. The regulated parties include refiners and importers fossil fuels and biofuels. The LCFS is a technology-neutral policy where regulated parties decide their mix of fuels to produce and/or to trade in order to reach the compliance targets (Table 7.1). The average CI of transportation fuels used in the California is dependent upon a variety of factors, including: (1) the GHG emissions from fuel production pathways, including emission from potential dLUC and iLUC for biofuels; (2) GHG emissions from fuel consumption based on existing vehicle technologies; (3) GHG reductions from advanced fuel manufacturing and vehicle technologies; (4) existing infrastructure (e.g., for fuel production and distribution), future investment and market penetration of advanced fuel production and vehicle technologies; (5) other factors such as the share of renewable energy for electricity generation which is relevant for market growth of electric vehicles as well as transportation fuel production. One likely reaction from the market in response to LCFS is the diversion of low-GHG fuels to California from other places in the US as a "quick solution". Such a "rationalization" of existing fuel production may lower the CI of transportation fuels used in California but may increase CI of transportation fuels used elsewhere, leading to no significant decrease of overall GHG emissions and possible delays in development of low-GHG technologies. In order to reduce the rationalization effect, LCFS sets less stringent compliance targets in early years to allow low-GHG technologies to develop (Farrell and Sperling 2007).

The direct GHG emissions from fuel production and use are modeled by GREET with customization of California-specific parameter values, such as feedstock supplies, transportation distances, vehicle emissions and electricity grid mix (Prabhu and Chowdhury 2017). The CA-GREET is enhanced with two sets of exogenous model inputs: (1) emissions from crude oil refining via OPGEE model and (2) emissions from dLUC and iLUC due to production and/or expansion of energy crops via GTAP model and AEZ-EF model.

7.5.2 Renewable Fuel Standard (RFS)

The Renewable Fuel Standard (RFS) requires that a certain volume of renewable fuels in total volume of transportation, heating or jet fuels in the US. There are four fuel categories that qualify as renewable fuel under the RFS: biomass-based diesel, cellulosic biofuel, advanced biofuel and total renewable fuel (typically ethanol derived from corn starch). Each fuel category has a minimum GHG reduction threshold compared to 2005 petroleum baseline, ranging from 20% (total renewable fuel) to 60% (cellulosic biofuel). Biodiesel typically qualify as both biomass-based diesel and advanced biofuel, dependent upon the feedstock used and energy consumption and GHG emissions during fuel production.

The major GHG emission sources considered under RFS include: domestic and international agricultural activities related to renewable fuel production, domestic and international land use (both direct and indirect) changes related to renewable fuel production, fuel production processes, transportation of feedstocks, materials and fuel distribution, and tailpipe emissions (www.epa.gov/renewable-fuel-standard-program). US EPA uses FASOM to evaluate the potential domestic dLUC and iLUC, and associated GHG emissions from increase in biofuel demand. For impacts on international level, US EPA uses the FAPRI-CARD model. The domestic N_2O emissions from agricultural soils are modeled by CENTURY model and its daily time-step version (DAYCENT). The emission factors for international countries are adopted from IPCC. The emissions associated with transportation of feedstocks, materials and fuel distribution are modeled using GREET model. CO_2 emissions from combusting biomass-based fuel are assumed to be carbon-neutral and thus are not included in the analysis. The CH_4 and N_2O emissions from tailpipes, on the other hand, are modeled using MOVES model (US EPA 2015). Energy consumption during fuel production is evaluated based on the data from both primary (e.g., industrial sources) and secondary (e.g., literature) data sources.

7.5.3 Renewable Energy Directive (RED) for EU

Renewable Energy Directive (RED) is established to promote the production and use of renewable energies in EU, with the goal of at least 20% of its total energy needs fulfilled by renewable sources by 2020. In addition, all EU member countries are required to have at least 10% of their transportation fuels from renewable sources by 2020 (European Commission 2009). RED also specifies "biofuels sustainability criteria" to ensure the production and use of biofuels in EU conform to the goal of reducing GHG emissions and protecting biodiversity (European Commission 2016). The criteria require the biofuels to reduce at least 50% of life cycle GHG emissions compared to the fossil fuels in 2017 and the reduction threshold is increased to 60% in 2018 for the biofuels produced from new production plants. Additional requirements include the prohibition of using raw materials from or converting land with high carbon stock (e.g., wetlands, forests) for biofuel production (European Commission 2016). RED also aims to reduce impacts from iLUC by limiting the use of biofuels from crops grown on agricultural land to 7% of total transportation fuel consumption in 2020.

The feedstocks for biodiesel that are regulated by the sustainability criteria include: rapeseed, soybean, sunflower, palm oil, waste cooking oil and animal fats. The soil N_2O emissions associated with feedstock growth are modeled by using IPCC tier 1 and tier 2 emission factors, and a tool named "GNOC" is developed for facilitating the calculation of soil N_2O emissions under different environmental and soil management conditions. CO_2 emissions from neutralization of acidity from nitrogen fertilizers is also included in the latest update of the RED. For potential iLUC, an extended version of MIRAGE (i.e., IFPRI-MIRAGE-BioF), a multi-sector, multi-country Computable General Equilibrium (CGE) model, is used to estimate the associated GHG emissions (European Commission 2012). The transportation of materials and fuel production processes are modeled based on the data representative of the EU market, which is sourced from international organizations, peer reviewed journal publications and reports, and primary data from stakeholders and industrial associations (European Commission 2016).

7.5.4 Challenges and Future Research on LCA for Biofuel Policies

Although LCA has been extensively used to estimate the potential environmental impacts of biofuels, its application in policy design is still nascent with several challenges to address including: (1) explicit modeling of spatiotemporal factors; and (2) interpreting and integrating LCA results of biofuels for gauging the outcome of policy instruments require additional modeling considerations that are typically alien to most LCA studies.

Existing biofuel policies predominantly focus on climate change while other impact categories, such as those related to water quality and biodiversity, should also be considered to avoid unintended consequences. Several guidelines, including ISO 13065 and Product Environmental Footprint (PEF) guide, have suggested a list of other impact categories (Bengtsson 2017). Nevertheless, many of these impact categories require location-specific modeling, thus including them in LCAs for biofuel policies without obscuring the spatial variations remain a challenge. Similar spatial variations also existing in LCI data. For example, the farming practices vary depending on the local climate and soil condition (Fast et al. 2012). Likewise, temporal variations also exist in LCI data and impact assessment. For example, the yield of perennial crops develops at different paces over time, therefore a simple annual average yield may not reflect the practice (McManus et al. 2015). These spatiotemporal factors, along with other factors (e.g., data quality and gap), constitute a large proportion of the inherent uncertainties in LCA of biofuels. A biofuel policy with single-value compliance goals (e.g., 50% reduction in GHG emissions compared to a baseline) is prone to neglecting these inherent variations and hence an uncertainty analysis is recommended to generate ranges of compliance goals (Fast et al. 2012; Koponen 2016).

Most existing studies either focus on a single biofuel life cycle with a normalized functional unit (e.g., 1 MJ) or assume the large-scale production and use of biofuels will not introduce any perturbation to the fuel market or the economy. Such a narrow-scope and static approach is also known as attributional LCA (ALCA). Although ALCA could provide high-resolution and detailed accounting of environmental impacts throughout the life cycle of biofuels, its limitation on capturing the indirect and scale effects (e.g., to sectors other than fuels) makes it inadequate for policy design biofuels which requires a holistic, multi-sector and dynamic analysis. For example, the co-product credits based on economic values may be subject to significant variations in real life, due to the factors such as market saturation. Another example of indirect and scale effects is the change in supply and demand elasticities of agricultural commodities in response to the change in demand of biofuels. A linear scaling of ALCA results may not account for such effects and therefore could generate misleading even erroneous policy recommendations. As a result, a consequence-oriented LCA approach, that is, consequential CLCA (CLCA), is considered to be more appropriate for policy design (Börjesson et al. 2010; Plevin et al. 2014). In addition to technological factors related to biofuels, CLCA aims to explicitly consider the following (nonexhaustive) list of factors that may affect the consequence of a biofuel policy: technology change during policy period (e.g., market diffusion of new technologies), input switch (e.g., switching fuel for manufacturing processes based on price), co-product credits, inter-sectoral competition for resources (e.g., dLUC/iLUC), demand-side changes (e.g., rebound effect of increased use of fossil energies when their prices are reduced due to the market expansion of biofuels), strategic behaviors (e.g., rationalization), selection of counterfactual scenarios (e.g., baseline scio-economic scenarios) and tradeoffs/synergies between specific instruments of the policy (Rajagopal and Zilberman 2013).

One major challenge in performing CLCA for biofuel policies is modeling iLUC and associated emissions. Given the complex nature of iLUC, one common approach is to use a computable general equilibrium (CGE) model (e.g., GTAP) or partial equilibrium (PE) model to estimate the reallocation of land resources in response to the market perturbation caused by a biofuel policy. A PE model typically covers a limited number of economic sectors (typically with a higher model resolution than CGE) that are of high relevance (e.g., agriculture, energy) while a CGE model covers the entire economy of a region or a country. Although there is currently no single agreed-upon framework for modeling emissions from iLUC, the Indirect Land Use Change Assessment Report from European Commission (European Commission 2012) and other studies (Koponen 2016; Bengtsson 2017) have indicated several groups of factors that may cause variations in modeling results and therefore should be harmonized through future research, including: model structure (e.g., PE vs. CGE) and coverage (e.g., geographical locations, policy time span), cause-effect relationships of economy (e.g., supply and demand elasticities), noneconomic drivers (e.g., change of dietary preferences), baseline scenarios (e.g., baseline land use scenarios), land use practices (e.g., soil emission modeling, yield development) and other CLCA related factors mentioned earlier.

References

Acero, Aitor P., Cristina Rodríguez, and Andreas Ciroth. 2015. "Impact Assessment Methods in Life Cycle Assessment and their Impact Categories". *OpenLCA*. Accessed on December 01, 2018. www.openlca.org/wp-content/uploads/2015/11/LCIA-METHODS-v.1.5.4.pdf.

Achten, Wouter M. J., Joana Almeida, Vincent Fobelets, Evelien Bolle, Erik Mathijs, Virendra P. Singh, Dina N. Tewari, Louis V. Verchot, and Bart Muys. 2010. "Life Cycle Assessment of Jatropha Biodiesel as Transportation Fuel in Rural India." *Applied Energy* 87 (12): 3652–60.

Achten, Wouter MJ, Erik Mathijs, Louis Verchot, Virendra P. Singh, Raf Aerts, and Bart Muys. 2007. "Jatropha Biodiesel Fueling Sustainability?" *Biofuels, Bioproducts and Biorefining* 1 (4): 283–91.

Acquaye, Adolf A., Thomas Wiedmann, Kuishang Feng, Robert H. Crawford, John Barrett, Johan Kuylenstierna, Aidan P. Duffy, S. C. Lenny Koh, and Simon McQueen-Mason. 2011. "Identification of 'Carbon Hot-Spots' and Quantification of GHG Intensities in the Biodiesel Supply Chain Using Hybrid LCA and Structural Path Analysis." *Environmental Science & Technology* 45 (6): 2471–78.

Adewale, Peter, Marie-Josée Dumont, and Michael Ngadi. 2015. "Recent Trends of Biodiesel Production from Animal Fat Wastes and Associated Production Techniques." *Renewable and Sustainable Energy Reviews* 45 (May): 574–88.

Asdrubali, Francesco, Giorgio Baldinelli, Francesco D'Alessandro, and Flavio Scrucca. 2015. "Life Cycle Assessment of Electricity Production from Renewable Energies: Review and Results Harmonization." *Renewable and Sustainable Energy Reviews* 42 (February): 1113–22.

Bailis, Robert E., and Jennifer E. Baka. 2010. "Greenhouse Gas Emissions and Land Use Change from Jatropha Curcas-Based Jet Fuel in Brazil." *Environmental Science & Technology* 44 (22): 8684–91.

Bengtsson, Jonas. "ARENA-Bioenergy-LCA-Literature-Review.Pdf." 2017. Accessed on November 19, 2018. https://arena.gov.au/assets/2017/02/ARENA-Bioenergy-LCA-Literature-Review.pdf.

Bilec, Melissa, Robert Ries, H. Scott Matthews, and Aurora L. Sharrard. 2006. "Example of a Hybrid Life-Cycle Assessment of Construction Processes." *Journal of Infrastructure Systems* 12 (4): 207–15.

Bonomi, Antonio, Otavio Cavalett, Lucas Pereira, Mateus Chaga, and Alexandre Souza. 2017. "Comparison of Biofuel Life CycleGHG Emissions Assessment Tools." Accessed on December 3. www.svebio.se/app/uploads/2017/05/Bonomi_Antonio_ABC17.pdf.

Börjesson, Pål, Linda Tufvesson, and Mikael Lantz. 2010. "Life Cycle Assessment of Biofuels in Sweden," (LUTFD2/TFEM–10/3061–SE + (1-88); Vol. 70). Lund University. Department of Technology and Society. Environmental and Energy Systems Studies.

Caldeira, Carla, João Queirós, Arash Noshadravan, and Fausto Freire. 2016. "Incorporating Uncertainty in the Life Cycle Assessment of Biodiesel from Waste Cooking Oil Addressing Different Collection Systems." *Resources, Conservation and Recycling* 112 (September): 83–92.

Canter, Christina E., Ryan Davis, Meltem Urgun-Demirtas, and Edward D. Frank. 2014. "Infrastructure Associated Emissions for Renewable Diesel Production from Microalgae." *Algal Research* 5 (July): 195–203.

Cerri, Carlos Eduardo Pellegrino, Xin You, Maurício Roberto Cherubin, Cindy Silva Moreira, Guilherme Silva Raucci, Bruno de Almeida Castigioni, Priscila Aparecida Alves, Domingos Guilherme Pellegrino Cerri, Francisco Fujita de Castro Mello, and Carlos Clemente Cerri. 2017. "Assessing the Greenhouse Gas Emissions of Brazilian Soybean Biodiesel Production." *PLOS ONE* 12 (5): e0176948.

Castanheira, Érica Geraldes, and Fausto Freire. 2017. "Environmental Life Cycle Assessment of Biodiesel Produced with Palm Oil from Colombia." *The International Journal of Life Cycle Assessment* 22 (4): 587–600.

Castanheira, Érica Geraldes, Renata Grisoli, Suani Coelho, Gil Anderi da Silva, and Fausto Freire. 2015. "Life-Cycle Assessment of Soybean-Based Biodiesel in Europe: Comparing Grain, Oil and Biodiesel Import from Brazil." *Journal of Cleaner Production* 102 (September): 188–201.

Chai, Ming, Qingshi Tu, Mingming Lu, and Jeffrey Yang. 2014. "Esterification Pretreatment of Free Fatty Acid in Biodiesel Production, from Laboratory to Industry." *Fuel Processing Technology* 125 (September): 106–13.

Chen, Rui, Zhangcai Qin, Jeongwoo Han, Michael Wang, Farzad Taheripour, Wallace Tyner, Don O'Connor, and James Duffield. 2018. "Life Cycle Energy and Greenhouse Gas Emission Effects of Biodiesel in the United States with Induced Land Use Change Impacts." *Bioresource Technology* 251 (March): 249–58.

Collet, Pierre, Arnaud Hélias, Laurent Lardon, Jean-Philippe Steyer, and Olivier Bernard. 2015. "Recommendations for Life Cycle Assessment of Algal Fuels." *Applied Energy* 154 (September): 1089–1102.

Corrado, Sara, Valentina Castellani, Luca Zampori, and Serenella Sala. 2018. "Systematic Analysis of Secondary Life Cycle Inventories When Modelling Agricultural Production: A Case Study for Arable Crops." *Journal of Cleaner Production* 172 (January): 3990–4000.

Curran, Mary Ann. 2016. *Goal and Scope Definition in Life Cycle Assessment*. New York, NY: Springer Berlin Heidelberg.

Davis, Ryan, Daniel Fishman, Edward D. Frank, Mark S. Wigmosta, Andy Aden, Andre M. Coleman, Philip T. Pienkos, Richard J. Skaggs, Erik R. Venteris, and Michael Q. Wang. 2012. "Renewable Diesel from Algal Lipids: An Integrated Baseline for Cost, Emissions, and Resource Potential from a Harmonized Model." ANL/ESD/12-4, PNNL-21437, 1044475.

Dolan, Stacey L., and Garvin A. Heath. 2012. "Life Cycle Greenhouse Gas Emissions of Utility-Scale Wind Power." *Journal of Industrial Ecology* 16 (s1): S136–54.

Dufour, Javier, and Diego Iribarren. 2012. "Life Cycle Assessment of Biodiesel Production from Free Fatty Acid-Rich Wastes." *Renewable Energy* 38 (1): 155–62.

European Commission. 2009. "Renewable Energy Directive - Energy - European Commission." *Energy*. Accessed on November 19, 2018. https://ec.europa.eu/energy/en/topics/renewable-energy/renewable-energy-directive

European Commission. 2016. "Sustainability Criteria - Energy - European Commission." *Energy*. Accessed on November 19, 2018. https://ec.europa.eu/energy/en/topics/renewable-energy/biofuels/sustainability-criteria.

European Commission. 2012. "ILUC Impact Assessment [SWD(2012) 343]." Accessed on November 20, 2018. https://ec.europa.eu/energy/sites/ener/files/swd_2012_0343_ia_en.pdf.

Farrell, Alexander E. (2007) "Low Carbon Fuel Standard for California-Part 1: Technical Analysis" Accessed on November 12, 2018. www.energy.ca.gov/low_carbon_fuel_standard/UC_LCFS_study_Part_1-FINAL.pdf.

Farrell, Alexander E., Richard J. Plevin, Brian T. Turner, Andrew D. Jones, Michael O'Hare, and Daniel M. Kammen. 2006. "Ethanol Can Contribute to Energy and Environmental Goals." *Science* 311 (5760): 506–8.

Farrell, Alexander E., and Daniel Sperling. 2007. "Low Carbon Fuel Standard for California-Part 2: Policy Analysis" Accessed on November 12, 2018. www.energy.ca.gov/low_carbon_fuel_standard/UC_LCFS_study_Part_2-FINAL.pdf.

Fast, Stewart, Mike Brklacich, and Marc Saner. 2012. "A Geography-Based Critique of New US Biofuels Regulations." *GCB Bioenergy* 4 (3): 243–52.

Firrisa, Melese Tesfaye, Iris van Duren, and Alexey Voinov. 2014. "Energy Efficiency for Rapeseed Biodiesel Production in Different Farming Systems." *Energy Efficiency* 7 (1): 79–95.

Fitzherbert, Emily B., Matthew J. Struebig, Alexandra Morel, Finn Danielsen, Carsten A. Brühl, Paul F. Donald, and Ben Phalan. 2008. "How Will Oil Palm Expansion Affect Biodiversity?" *Trends in Ecology & Evolution* 23 (10): 538–45.

Handler, Robert M., Christina E. Canter, Tom N. Kalnes, F. Stephen Lupton, Oybek Kholiqov, David R. Shonnard, and Paul Blowers. 2012. "Evaluation of Environmental Impacts from Microalgae Cultivation in Open-Air Raceway

Ponds: Analysis of the Prior Literature and Investigation of Wide Variance in Predicted Impacts." *Algal Research* 1 (1): 83–92.

Hansen, Sune Balle, Stig Irving Olsen, and Zaini Ujang. 2014. "Carbon Balance Impacts of Land Use Changes Related to the Life Cycle of Malaysian Palm Oil-Derived Biodiesel." *The International Journal of Life Cycle Assessment* 19 (3): 558–66.

Harsono, Soni Sisbudi, Annette Prochnow, Philipp Grundmann, Anja Hansen, and Claudia Hallmann. 2012. "Energy Balances and Greenhouse Gas Emissions of Palm Oil Biodiesel in Indonesia." *GCB Bioenergy* 4 (2): 213–28.

Hendrickson, Chris T., Arpad Horvath, Satish Joshi, Markus Klausner, Lester B. Lave, and Francis C. McMichael. 1997. "Comparing Two Life Cycle Assessment Approaches: A Process Model vs. Economic Input-Output-Based Assessment." In *Proceedings of the 1997 IEEE International Symposium on Electronics and the Environment. ISEE-1997*, 176–81. doi:10.1109/ISEE.1997.605313.

Hertwich, Edgar, Niko Heeren, Brandon Kuczenski, Guillaume Majeau-Bettez, Rupert J. Myers, Stefan Pauliuk, Konstantin Stadler, and Reid Lifset. 2018. "Nullius in Verba1: Advancing Data Transparency in Industrial Ecology." *Journal of Industrial Ecology* 22 (1): 6–17.

Hums, Megan E., Richard A. Cairncross, and Sabrina Spatari. 2016. "Life-Cycle Assessment of Biodiesel Produced from Grease Trap Waste." *Environmental Science & Technology* 50 (5): 2718–26.

Kim, Dong-Shik, Mohammadmatin Hanifzadeh, and Ashok Kumar. 2018. "Trend of Biodiesel Feedstock and Its Impact on Biodiesel Emission Characteristics." *Environmental Progress & Sustainable Energy* 37 (1): 7–19.

Kim, Hyung Chul, Vasilis Fthenakis, Jun-Ki Choi, and Damon E. Turney. 2012. "Life Cycle Greenhouse Gas Emissions of Thin-Film Photovoltaic Electricity Generation." *Journal of Industrial Ecology* 16 (s1): S110–121.

Kiss, Ferenc E., Radoslav D. Micic, Milan D. Tomić, Emilija B. Nikolić-Djorić, and Mirko Đ. Simikić. 2014. "Supercritical Transesterification: Impact of Different Types of Alcohol on Biodiesel Yield and LCA Results." *The Journal of Supercritical Fluids* 86 (February): 23–32.

Kiwjaroun, Choosak, Chanporn Tubtimdee, and Pornpote Piumsomboon. 2009. "LCA Studies Comparing Biodiesel Synthesized by Conventional and Supercritical Methanol Methods." *Journal of Cleaner Production* 17 (2): 143–53.

Koponen, Kati. 2016. "Challenges of an LCA Based Decision Making Framework – The Case of EU Sustainability Criteria for Biofuels," PhD diss., Aalto University. Aalto University publication series (DOCTORAL DISSERTATIONS 85/2016)

Kumar, Sunil, Jasvinder Singh, S. M. Nanoti, and M. O. Garg. 2012. "A Comprehensive Life Cycle Assessment (LCA) of Jatropha Biodiesel Production in India." *Bioresource Technology* 110 (April): 723–29.

Kwan, Thomas A., Qingshi Tu, and Julie B. Zimmerman. 2016. "Simultaneous Extraction, Fractionation, and Enrichment of Microalgal Triacylglyerides by Exploiting the Tunability of Neat Supercritical Carbon Dioxide." *ACS Sustainable Chemistry & Engineering* 4 (11): 6222–30.

Leow, Shijie, Brian D. Shoener, Yalin Li, Jennifer L. DeBellis, Jennifer Markham, Ryan Davis, Lieve M. L. Laurens, et al. 2018. "A Unified Modeling Framework to Advance Biofuel Production from Microalgae." *Environmental Science & Technology* 52 (22): 13591–99.

Lesage, Pascal, and Réjean Samson. 2016. "The Quebec Life Cycle Inventory Database Project." *The International Journal of Life Cycle Assessment* 21 (9): 1282–89.

Life cycle assessment - Requirements and guidelines. ISO 14044, 2006.

Liu, Xiaowei, Andres F. Clarens, and Lisa M. Colosi. 2012. "Algae Biodiesel Has Potential despite Inconclusive Results to Date." *Bioresource Technology* 104 (January): 803–6.

López, Dora E., Joseph C. Mullins, and David A. Bruce. 2010. "Energy Life Cycle Assessment for the Production of Biodiesel from Rendered Lipids in the United States." *Industrial & Engineering Chemistry Research* 49 (5): 2419–32.

Malça, João, and Fausto Freire. 2011. "Life-Cycle Studies of Biodiesel in Europe: A Review Addressing the Variability of Results and Modeling Issues." *Renewable and Sustainable Energy Reviews* 15 (1): 338–51.

McManus, Marcelle C., Caroline M. Taylor, Alison Mohr, Carly Whittaker, Corinne D. Scown, Aiduan Li Borrion, Neryssa J. Glithero, and Yao Yin. 2015. "Challenge Clusters Facing LCA in Environmental Decision-Making—What We Can Learn from Biofuels." *The International Journal of Life Cycle Assessment* 20: 1399–414.

Mehta, Pramod S., and K. Anand. 2009. "Estimation of a Lower Heating Value of Vegetable Oil and Biodiesel Fuel." *Energy & Fuels* 23 (8): 3893–98.

Menten, Fabio, Benoît Chèze, Laure Patouillard, and Frédérique Bouvart. 2013. "A Review of LCA Greenhouse Gas Emissions Results for Advanced Biofuels: The Use of Meta-Regression Analysis." *Renewable and Sustainable Energy Reviews* 26 (October): 108–34.

Montefrio, Marvin Joseph, Tai Xinwen, and Jeffrey Philip Obbard. 2010. "Recovery and Pre-Treatment of Fats, Oil and Grease from Grease Interceptors for Biodiesel Production." *Applied Energy* 87 (10): 3155–61.

Morais, Sérgio, Teresa M. Mata, António A. Martins, Gilberto A. Pinto, and Carlos A. V. Costa. 2010. "Simulation and Life Cycle Assessment of Process Design Alternatives for Biodiesel Production from Waste Vegetable Oils." *Journal of Cleaner Production* 18 (13): 1251–59.

Nelson, Richard G., and Mark D. Schrock. 2006. "Energetic and Economic Feasibility Associated with the Production, Processing, and Conversion of Beef Tallow to a Substitute Diesel Fuel." *Biomass and Bioenergy* 30 (6): 584–91.

Piastrellini, Roxana, Alejandro Pablo Arena, and Bárbara Civit. 2017. "Energy Life-Cycle Analysis of Soybean Biodiesel: Effects of Tillage and Water Management." *Energy* 126 (May): 13–20.

Plevin, Richard J., Mark A. Delucchi, and Felix Creutzig. 2014. "Using Attributional Life Cycle Assessment to Estimate Climate-Change Mitigation Benefits Misleads Policy Makers." *Journal of Industrial Ecology* 18 (1): 73–83.

Portugal-Pereira, Joana, Jun Nakatani, Kiyo Kurisu, and Keisuke Hanaki. 2016. "Life Cycle Assessment of Conventional and Optimised Jatropha Biodiesel Fuels." *Renewable Energy* 86 (February): 585–93.

Prabhu, Anil, and Hafizur Chowdhury. 2017. "CA-GREET 3.0 Supplemental Document and Tables of Changes,". Accessed on November 12, 2018. www.arb.ca.gov/fuels/lcfs/lcfs_meetings/11062017greet_supp.pdf.

Pradhan, A., D. S. Shrestha, A. McAloon, W. Yee, M. Haas, and J. A. Duffield. 2011. "Energy Life-Cycle Assessment of Soybean Biodiesel Revisited." *Transactions of the ASABE* 54 (3): 1031–39.

Quek, Augustine, and Rajasekhar Balasubramanian. 2014. "Life Cycle Assessment of Energy and Energy Carriers from Waste Matter – A Review." *Journal of Cleaner Production* 79 (September): 18–31.

Rajagopal, Deepak, and David Zilberman. 2010. "On Environmental Lifecycle Assessment for Policy Selection." Accessed on November 12, 2018. https://escholarship.org/content/qt39k9h976/qt39k9h976.pdf

Rajagopal, Deepak, and David Zilberman. 2013. "On Market-Mediated Emissions and Regulations on Life Cycle Emissions." *Ecological Economics* 90 (June): 77–84.

Reuters. "EU to Phase out Palm Oil from Transport Fuel by 2030," Accessed on June 14, 2018. https://af.reuters.com/article/commoditiesNews/idAFL8N1TG4J1.

Sandia National Laboratory. 2013. "Better Monitoring and Diagnostics Tackle Algae Biofuel Pond Crash Problem – Sandia Energy." Accessed on December 9, 2018. https://energy.sandia.gov/better-monitoring-and-diagnostics-tackle-algae-biofuel-pond-crash-problem/.

Sheehan, John, Vince Camobreco, James Duffield, Michael Graboski, Michael Graboski, and Housein Shapouri. 1998. "Life Cycle Inventory of Biodiesel and Petroleum Diesel for Use in an Urban Bus." NREL/SR–580-24089, 1218369.

Shuba, Eyasu Shumbulo, and Demeke Kifle. 2018. "Microalgae to Biofuels: 'Promising' Alternative and Renewable Energy, Review." *Renewable and Sustainable Energy Reviews* 81 (January): 743–55.

Sills, Deborah L., Vidia Paramita, Michael J. Franke, Michael C. Johnson, Tal M. Akabas, Charles H. Greene, and Jefferson W. Tester. 2013. "Quantitative Uncertainty Analysis of Life Cycle Assessment for Algal Biofuel Production." *Environmental Science & Technology* 47 (2): 687–94.

Subhadra, Bobban G., and Mark Edwards. 2011. "Coproduct Market Analysis and Water Footprint of Simulated Commercial Algal Biorefineries." *Applied Energy*, Special Issue of Energy from algae: Current status and future trends 88 (10): 3515–23.

Talens Peiró, L., L. Lombardi, G. Villalba Méndez, and X. Gabarrell i Durany. 2010. "Life Cycle Assessment (LCA) and Exergetic Life Cycle Assessment (ELCA) of the Production of Biodiesel from Used Cooking Oil (UCO)." *Energy*, ECOS 2008, 35 (2): 889–93.

Thamsiriroj, T., and J. D. Murphy. 2009. "Is It Better to Import Palm Oil from Thailand to Produce Biodiesel in Ireland than to Produce Biodiesel from Indigenous Irish Rape Seed?" *Applied Energy* 86 (5): 595–604.

Tropical Forest Alliance. 2017. "Tropical Forest Alliance 2020 Africa Palm Oil Inititative." Accessed on December 14, 2018. www.tfa2020.org/wp-content/uploads/2017/10/BN5_English_Final_5Oct.pdf.

Tu, Qingshi, Matthew Eckelman, and Julie Zimmerman. 2017. "Meta-Analysis and Harmonization of Life Cycle Assessment Studies for Algae Biofuels." *Environmental Science & Technology* 51 (17): 9419–32.

Tu, Qingshi, Matthew Eckelman, and Julie Beth Zimmerman. 2018. "Harmonized Algal Biofuel Life Cycle Assessment Studies Enable Direct Process Train Comparison." *Applied Energy* 224 (August): 494–509.

Tu, Qingshi, Mingming Lu, Worrarat Thiansathit, and Tim C. Keener. 2016. "Review of Water Consumption and Water Conservation Technologies in the Algal Biofuel Process." *Water Environment Research* 88 (1): 21–8.

Tu, Qingshi, and Bryant E. McDonnell. 2016. "Monte Carlo Analysis of Life Cycle Energy Consumption and Greenhouse Gas (GHG) Emission for Biodiesel Production from Trap Grease." *Journal of Cleaner Production* 112 (January): 2674–83.

Tu, Qingshi, Jingjing Wang, Mingming Lu, Andrew Brougham, and Ting Lu. 2016. "A Solvent-Free Approach to Extract the Lipid Fraction from Sewer Grease for Biodiesel Production." *Waste Management* 54 (August): 126–30.

Tuntiwiwattanapun, Nattapong, Parnuwat Usapein, and Chantra Tongcumpou. 2017. "The Energy Usage and Environmental Impact Assessment of Spent Coffee Grounds Biodiesel Production by an In-Situ Transesterification Process." *Energy for Sustainable Development* 40 (October): 50–8.

United Nations Development Programme. "Sustainable Palm Oil Initiative." Accessed on December 14, 2018. www.undp.org/content/dam/undp/library/Environment%20and%20Energy/Green%20Commodities%20Programme/Indonesia%20Factsheet%20pdf.pdf.

US EPA. 2015. "Regulatory Impact Analysis-Renewable Fuel Standard Program." Accessed on November 12, 2018. www.epa.gov/sites/production/files/2015-08/documents/420r07004.pdf

US EPA, OA. 2016. "Understanding Global Warming Potentials." Overviews and Factsheets. *US EPA*. Accessed on January 12. www.epa.gov/ghgemissions/understanding-global-warming-potentials.

van Duren, Iris, Alexey Voinov, Oludunsin Arodudu, and Melese Tesfaye Firrisa. 2015. "Where to Produce Rapeseed Biodiesel and Why? Mapping European Rapeseed Energy Efficiency." *Renewable Energy* 74 (February): 49–59.

Vijay, Varsha, Stuart L. Pimm, Clinton N. Jenkins, and Sharon J. Smith. 2016. "The Impacts of Oil Palm on Recent Deforestation and Biodiversity Loss." *PLOS ONE* 11 (7): e0159668.

Ward, Hauke, Leonie Wenz, Jan C. Steckel, and Jan C. Minx. 2018. "Truncation Error Estimates in Process Life Cycle Assessment Using Input-Output Analysis: Truncation Error Estimates in Life Cycle Assessment." *Journal of Industrial Ecology* 22 (5): 1080–91.

Wood, Richard, Konstantin Stadler, Tatyana Bulavskaya, Stephan Lutter, Stefan Giljum, Arjan de Koning, Jeroen Kuenen, et al. 2015. "Global Sustainability Accounting—Developing EXIOBASE for Multi-Regional Footprint Analysis." *Sustainability* 7 (1): 138–63.

Yang, Jia, Ming Xu, Xuezhi Zhang, Qiang Hu, Milton Sommerfeld, and Yongsheng Chen. 2011. "Life-Cycle Analysis on Biodiesel Production from Microalgae: Water Footprint and Nutrients Balance." *Bioresource Technology*, Special Issue: Biofuels - II: Algal Biofuels and Microbial Fuel Cells 102 (1): 159–65.

Yang, Yi, Reinout Heijungs, and Miguel Brandão. 2017a. "Hybrid Life Cycle Assessment (LCA) Does Not Necessarily Yield More Accurate Results than Process-Based LCA." *Journal of Cleaner Production* 150 (May): 237–42.

Yang, Yi, Wesley W. Ingwersen, Troy R. Hawkins, Michael Srocka, and David E. Meyer. 2017b. "USEEIO: A New and Transparent United States Environmentally-Extended Input-Output Model." *Journal of Cleaner Production* 158 (August): 308–18.

Yee, Kian Fei, Kok Tat Tan, Ahmad Zuhairi Abdullah, and Keat Teong Lee. 2009. "Life Cycle Assessment of Palm Biodiesel: Revealing Facts and Benefits for Sustainability." *Applied Energy*, Bio-fuels in Asia 86 (November): S189–96.

8

Future Direction of Biodiesel in the Next Decade

Dong-Shik Kim and Kaushik Shandilya

The global recession in the last decade appeared to be the most significant factor that determined the current status and the future of renewable energies. Currently, renewable energy industries in almost all the countries (except a few such as Indonesia and Brazil) are undergoing downsizing or shutting down due to the reduced government supports, which results in weaker competitiveness in the energy market. However, in many countries including Canada and US, the notion that renewable energy can help boost job creation and be beneficial to economy is gaining more popularity, and for that reason, ironically the past economic downturn may be a great opportunity for biodiesel (BD) to be more competitive in the energy market. Figure 8.1 shows a competitive increase of renewable energy consumption until 2040 compared to other energy forms.

8.1 Directions in Countries

a. Asia

- According to the literature, Korea planned for implementation of B3 in 2012. Renewable portfolio standards for South Korea replaced the previous feed-in tariff system in 2012 and require South Korea's major electric utilities to gradually increase the renewable energy share in their power generation portfolios to an average of 10% by 2024.
- Thailand planned for B10 in 2012. Thailand has very aggressive plans for BD, whereas they were projecting 8.5 million liters of BD production in 2012. Thailand has also emphasized a growing share of renewables to contribute to the country's long-term power generation.
- China is targeting 2 billion tons of BD to be produced by 2020. China, now the world's leading investor in the renewable energy sector, will produce at least 15% of overall energy consumption by 2020 from renewable sources.

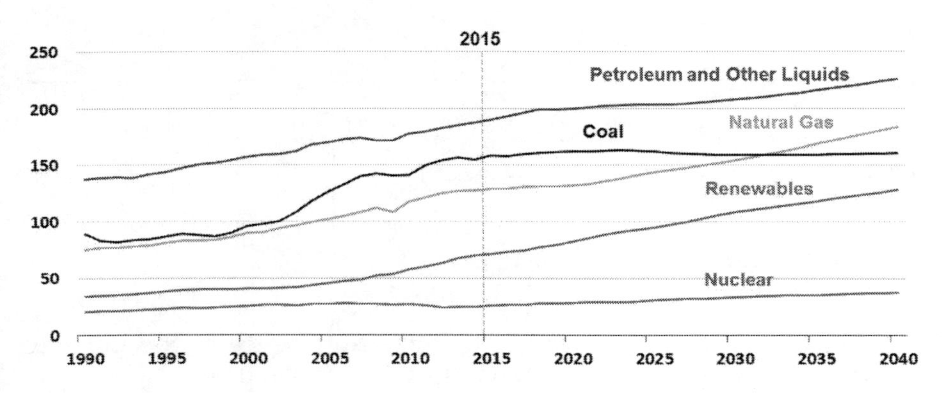

FIGURE 8.1
World energy consumption by fuel in quadrillion Btu. (EIA. International Energy Outlook Executive Summary, September 2017. www.eia.gov/outlooks/ieo/pdf/exec_summ.pdf).

- Japan: As a part of the revised energy policy plan, Japan is trying to encourage increased use of renewable energy for power generation from renewable sources. Renewable energy, apart from hydroelectricity, made up slightly more than 3% of Japan's total energy consumption and about 8% (79 TWh) of the country's total electricity generation in 2015. Japan wants to boost generation from both hydroelectricity and other renewables to 22–24% by 2030.

b. **South America**
- Argentina: On April 2016, the Ministry of Energy and Mining introduced a resolution raising the required blending mandate for ethanol in regular gasoline from 10 to 12%.
- Bolivia: Bolivia planned for B20 in 2015.
- Brazil: Brazil planned for B5 in 2013. In 2016, the country produced approximately 65,000 b/d of BD. In March 2016, Brazil increased the BD-use mandate from B7 to B10 by 2019, with the following implementation timeline: B8, B9 and B10 by March 2017, March 2018, and March 2019, respectively. Brazil is also the second-largest producer (494,000 b/d in 2016, a 3.5% decrease from 2015) and consumer of ethanol in the world after the US.
- Dominican Republic planned for B2 in 2013.
- Uruguay: Uruguay planned for B5 in 2012.

c. Europe
- Germany: In 2016, Germany was one of the largest BD producers in the world with more than 3 billion liters of production per year.

Germany is planning to increase their production of BD to more than 3.5 billion liters in 2018.

- The UK planned to increase BD consumption to 348,720 ton in 2015. The UK was not among one of the top five producing EU Member States (Germany, France, Netherlands, Spain, and Poland). The UK's rank in BD production was dropped from seventh place in 2015 (648 million liters) to tenth in 2017 (420 million liters), due to imports from other Member States, which were more competitive than domestic production.
- France is the second largest BD producer among EU countries. France is targeting to increase its production to about 2,390 million liters in 2017.
- Italy plans to increase BD production gradually. It is estimated that Italy is aiming for an increase of production to 665 million liters in 2017.
- The Netherlands is also targeting to produce 665 million liters of BD in 2017.
- Spain is targeting to increase their BD production to 1,080 million liters in 2017.
- Belgium is targeting 570 million liters production of BD in 2017.

According to a report from the International Energy Agency (IEA), it was estimated that sustainable global biomass production could reach between 200 and 400 EJ (1 EJ = 10^{18} J) in the next century with others suggesting even higher values. These figures are regarded substantial considering the fact that 2007 fossil fuel energy use was 388 EJ (IEA Bioenergy, 2007) [1]. It is quite likely that with further advances in both new feedstock supplies such as algae and improved conversion technologies, second generation biofuels will play a key role in the transport fuel mix.

Second generation conversion technologies like diesel synthesized via Fischer-Tropsch process or gasification methods are under intensive R&D to increase their viability (Digman et al., 2009 [2]; Ail and Dasappa, 2016 [3]). Typically, these technologies will require large plant capacities to achieve economies of scale. Large plants imply procuring biomass from a wide area – a logistical and economical challenge. Using calculations based on arable land, it is technically possible that by 2050 production figure could rise sustainably and not in competition with food to 11%. For a detailed explanation of these values, refer to Doornbosch and Steenblik (2007) [4].

EU made a common agricultural policy in 2008 (See Table 2.2 in Chapter 2 of this book). The only element measure was the abolishment (in 2010) of the energy crop premium of 45 Euro per hectare, which applied to rapeseed grown for BD production. Although the imports of BD from the US faced temporary antidumping and antisubsidy duties in March 2009 and tariffs ranging from 26 to 41 Euro per kilogram were applied for an initial period of 6 months, surging imports of BD produced in the US and reducing benefits from the government supports were considered to severely injure the competitiveness of the EU BD industry.

Situation in Germany in the last decade was not any better than the rest of EU. In February, 2009, the policy for BD taxation was implemented. This policy was in line with the overall policy to gradually scale back the support provided to the BD industry from January 2009 onward. Taxation of vegetable oils used as BD feedstock will increase from 0.1 to 0.18 Euros per liter while the tax for B100 was raised from 0.15 to 0.18 Euro per liter. In March 2009, the EU commission requested Germany that the sustainability criteria included in the national law on the promotion of biofuels should be streamlined with the relevant EU directive. As a result, imports of vegetable oils – especially soy and palm oil for local BD production – are expected to resume. Furthermore, new legislation, excluding biofuels that had previously received state aids from being eligible for national incentives and mandates, went ahead to pass. Consequently, the import of BD that has benefitted from the state aid could become uneconomical.

In March 2009, France made a policy on its BD taxation very similar to Germany. The taxes on BD and its feedstock were raised in January. France announced that all tax advantages granted to BD would be discontinued by 2012, and as a result, the amount of tax discount for BD 100 reduced to 5.14 Euro/liter in 2012 from 17.74 Euro/liter in 2011. The tax discount on BD 100 was further reduced to 2.14 Euro/liter since 2013. The tax discount for BD blends was cut to zero since 2007 (See Table 2.2 in Chapter 2 of this book for more details).

In an attempt to curb down the BD imports from Argentina and Indonesia to Europe, the EC enforced antidumping duties (AD) on BD imports from these origins as of May 29, 2013. As a result, imports from both countries have dropped considerably in 2013 and almost ceased in 2014. The void was partially filled with domestic EU production and partially with higher imports from countries not covered by AD (Malaysia, South Korea, India, and Brazil).

In 2015, about 509 million liters of BD was imported to EU. The majority of BD imports occurred through the Netherlands, Spain, and Bulgaria. BD imports are constrained by the sustainability requirements laid down in the Renewable Energy Directive (RED). Since April 1, 2013, all biofuels must achieve greenhouse gas (GHG) savings of at least 35%. Default values of BD produced from both soybean oil and palm oil are set lower than that in the RED.

EU BD exports to destinations outside the bloc are marginal and normally only amount to around one percent of production. The exceptional increase of exports in 2013 was due to higher exports to the US. In 2015, the top three export destinations were Norway, Switzerland, and the US receiving 68, 28, and 6% of EU exports, respectively.

The government of India set an indicative target of a minimum 20% blended diesel across the country by 2017. The policy also suggested removing all central taxes on BD and, accordingly, declaring good status that would ensure a uniform 4% sales tax on the product across the state. As for the policy, a certification mechanism would be put in place for blending exercise that would have to conform to Bureau of Indian Standards specifications. India has no separate BD policy. BD production is and will be taken up by nonedible oil seeds produced from waste, degraded, or marginal lands. The focus would be on indigenous production of BD feedstock and therefore, an import of free fatty acid (FFA) waste oil, such as palm oil, would not be permitted. The national BD policy carries messages that BD may be brought under the ambit of declared goods by the government to ensure unrestricted movement within and outside the states. It is also stated in the policy that no taxes and duties should be levied on BD (http://petroleum.nic.im/biodiesel.pdf) [5]. NGOs and private businesses will be implementing the policy. Major focus has been and will be on fuel security. The environmental impact of the policy suggests that it will improve the environmental quality without compromising the fuel security.

8.2 Directions of Companies

In general, government policies have been developed for regulatory, economic, and informational components, but the technological component has been absent. For example, Indian biofuel policies specified administrative supports for building up production capacities, but no policies for financial and technological support were promulgated. Therefore, it was regarded that broader economic impact of biodiesel production would be creating green jobs, but the policies did not contribute socially. Furthermore, environmentally, the policy focused on reducing air pollution, but did not focus on affected areas with noise pollution. The policy turned out to cost to private sectors and to households. The Indian government's role in this policy was negligible, and eventually this policy was disapproved by the general public. Feasibility analysis suggested also that the National Institution for Transforming India (NITI) Commission (former Planning Commission of India) restricted the implementation of the policy in India.

In Brazil, Petrobras invested $0.48 billion on BD from 2009 to 2013. They aimed to produce 640 million liters of BD in Brazil until 2013. Since January 2010, all diesel fuel sold in Brazil is required to have at least 5% biodiesel. Petrobras supplied 20.2% of Brazil's biodiesel and acted as a market catalyst by securing and blending biodiesel supplies and furnishing these to smaller distributors as well as its own service stations. Petrobras owned three biodiesel plants and through its 50% interest in BSBIOS EnergiaRenovável S.A. (BSBIOS), it owned two additional plants. In 2013, the biodiesel production capacity of these five plants totaled 760 million liters that outpaced the originally planned. In 2017, the company reported that it supplied 16% of Brazil's biodiesel (assuming 100% of BSBIOS Indústria e Comércio de Biodiesel Sul Brasil S.A. (BSBIOS Sul Brasil) production). Petrobras directly owns three biodiesel plants, but Quixadá biodiesel plant had its own operation stopped in November 2016 due to weak economic results and it is in restorative hibernation state and, through its 50% interest in BSBIOS Sul Brasil, it still owns two additional plants. The biodiesel production capacity of these five plants in 2017 totals 1 billion liters/year.

Chevron has Galveston Bay BD LP plant. The initial production was 20 million gallons of BD per year and operations expanded to produce 110 million gallons per year by 2010. This plant produces BD from soybean and other renewable feedstock. Their goal of the 5-year alliance with Georgia Tech was to develop commercially viable processes for making transportation fuels from renewable resources such as agricultural waste. Bioselect fuels LLC, a division of Standard Renewable Energy Group, LLC and Chevron Technology Ventures LLC (CVT) started Galveston Bay BD project in 2006. Chevron Technology Venture subsidiary of Chevron USA has 22% equity stakes in $25 million BD refineries. In 2008 Chevron withhold investment in Galveston Bay BD project. Chevron Algy CRADA project focuses on liquid transportation fuel.

Total is Europe's leader in first generation biofuels and they use rapeseed oil and sunflower oil to make BD. Their BD is called BOME. Total was the largest reseller of BD until 2002. Total signed a MOU with NESTE oil, a Finish oil company, in 2005 for a joint study on the possibilities of building a production plant for a new generation BD product. Total uses hydrotreating of vegetable oil and animal fat to produce a high-performance BD.

Eni produces BD using rapeseed. They sold 210 kilo tonnes of BD in 2007. They are investing $418 million in 4 new BD facilities. Petrobras and Eni S.P.A, Italy, signed a renewal of MOU in 2007 to develop and commercialize technologies and build 4 BD plants in Brazil.

BD from palm oil shows notable progress when Malaysian Palm Oil Board and Petronas have developed a BD production technology on the basis of palm oil first time, of which the main purpose was production of carotene, vitamin Q, lipids, and other substances, where BD was produced as a byproduct and exported. Toyota, Nippon Oil, and Petronas started a joint development study in 2007 and began producing palm oil–based BD in 2009 for test basis in Malaysia. For this study, Petronas supplied palm oil and Nippon oil developed a refining technology to convert it to BD for automobiles. Toyota checked if the developed fuels are safe to use for car engines. Finally, Toyota and Nippon oil declined to confirm the BD joint development.

With soybean as a feedstock for BD, White Gate refinery in Cork, Ireland, produced 42,000 gallons of BD per day from soybean. Borger refinery started BD production from by-product animal fat through alliance with Tyson Foods in 2007. Their production capacity is 60,000 tonnes per year. Conoco Philips is generating first generation BD from soybean oil.

Major oil companies in US have sped up BD business. ExxonMobile started mixing petroleum derived diesel and BD fuels together at its first BD blending project in the US in 2008. They are focusing on the next generation BD. BP's Bulwer refinery in Queensland, Australia produced around 110 million liters of BD per year from tallow feedstock. BP is funding a $9.4 million project through the Energy and Resources institute (TERRY) of the Indian state of Andra Prudesh (AP) to demonstrate the feasibility of BD from Jatropha Curcas, a nonedible, all appearing crop. This study would take 10 years and planned to produce 9 million liters of BD per year. BP made a strategy in 2007 to produce feedstock that do not compete with food production and can be grown sustainably on fallow land. This project was implemented to cultivating Jatropha on around 1 million hectares in Southern Africa, India, Southeast Asia, and Central and South America.

Neste Oil's synthetic NExBTL BD was the world's first second generation BD to be launched commercially. This was flexibly manufactured from plant oil or animal fat using a proprietary conversion process for vegetable oils and animal fats. It is produced by direct catalytic hydrogenation of plant oil into the corresponding alkane. The company commercially launched a unique renewable diesel that contained the BTL (biomass-to-liquid) BD in Finland in 2008. The BD plant at Porvoo refinery in Finland has production capacity of 170,000 tonnes per year using 100% renewable raw materials in the form of vegetable oils and animal fats. The company invested approximately 100 million Euros in the new facility. Neste Oil set up another BD plant at Porvoo refinery, Finland, in 2008 with the same capacity. Neste Oil has announced that they intend to construct BD plants in Singapore and Rotterdam in the Netherlands at a cost of 440 million Euros and 670 million Euros. The maximum production capacity reached 800,000 tonnes per year of each. The Singapore plant was started in 2010. Neste Oil made an alliance with Air Liquide to supply hydrogen to their BD plant. They make the same kind of alliance with SOXARL. Neste Oil signed MOU to evaluate possibilities to jointly build a large scale production plant for next generation BD fuel in Europe. Neste Oil also made an alliance with Austran oil and gas group, OMV to build a BD plant near Vienna, Austria. Neste Oil's BD project focused on using Fishcher-Tropsch process to produce crude BD from gas. This project is a 50/50 partnership between Neste oil and Stora ENSO.

The Philippines has a highly developed plan for BD. Its policy requires that the government vehicles use BD product that comply with the Philippines National Standard (PNS, 2020, 2003). By blending a DOE accredited BD product with the diesel to comply with 1% coconut methyl ester for use in diesel fad vehicles (www.ops.gov.ph/records/ao_no103.htm) [6]. The Biofuel act of 2006 called for 5% BD blending by 2011. The government provided no specific tax exemption in the VAT for raw materials and production related to BD. The law aimed to reduce dependence on imported fuel with due regard to the protection of public health, the environment, and natural ecosystems consistent with the country's sustainable economic growth, which would benefit livelihood by mandating the use of BD. The law outlines the goals as follow:

a. To develop indigenous renewable and sustainableclean energy sources to reduce dependence on imported oil.
b. To mitigate toxic and greenhouse gas emissions.
c. To increase rural employment and income.
d. To ensure the availability of alternative and renewable clean energy without any detriment to the natural ecosystem, biodiversity, and fuel reserves of the country.

Brazilian biofuel program is pro-BD. Their national program of production and use of BD was implemented in 2005. The program aimed to encourage small producers and farmers from least developed regions in Brazil to get involved with BD production. Through this program, the use of BD blending up to 2% was authorized according to product availability, as well as the establishment of mandatory blending of 2% of BD with the diesel sold throughout the country from January 2008 aiming to reach 5% in 2013. In 2017, the mandatory biodiesel blending went up to 8%.

In December 2008, Thai Government started supporting palm oil-based BD production. The government plans to purchase palm oil from oil mills at a guaranteed price, hoping to prevent prices from falling further. In Indonesia, In February 2009, the government subsidized the sale of palm oil-based diesel with low fossil fuel prices that had compromised the profitability of BD production. In March 2009, the government subsidized the sale of biofuels by state owned companies depending on how the prices for fossil fuel and biofuel feedstock, notably palm oil, develop. The subsidy would amount to 1,000 rupees ($0.07) per liter of biofuel distributed.

Palm oil-based BD should account for about three-quarters of all biofuels saved. In Malaysia, in March 2009, the government confirmed that blending of diesel fuel with 5% palm oil biodiesel became mandatory on January 2010. In 2009, all government vehicles started using the blend. To remain competitive, BD fuels continued to get benefits from a subsidy. Once fully operational, a new program is expected to absorb 500,000 tonnes of BD annually (Lim and Teong, 2010) [7].

Waste-to-Energy (WTE) is a new market that is growing rapidly and reached nearly $13.6 billion in 2016. Private companies with the help of local municipalities in the US and Canada are building several plants to

TABLE 8.1

Long-term (national) goals in energy policy

Challenger	Target(s)
World	Reduction in GHG emissions by 5.2% on 1990 levels throughout the 2008–2012 period
Germany	Biodiesel target of 10% by 2015 Reduction of GHG emissions by 40% by 2020 against 1990 levels
China	Share of 10% renewable energy by 2010, 15% by 2020
UK	Renewable transport fuels accounting for 5% in 2010 and 10% in 2015 Reduction in CO_2 emissions by 26–32% (2020) to 80% (by 2050) against 1990 baseline
USA	Soya biodiesel share of 4% in 2016 Replacement of 15% of current gasoline consumption by 2017
India	Share of 10% renewable energy by 2012

(Source: Bart et al., 2010) [8]

produce BD from waste. Waste Management in the US has stated that the trash they collect could produce $50 billion of BD, jet fuel, and other useful products after all the technologies are fully developed and implemented. Table 8.1 show the long-term goals of a few selective countries in their energy policies. These countries aim to replace 5–15% fossil fuel with renewable energy by 2017–2020.

As stated earlier, in 2012 Korea has launched an ambitious short- and long-term plans to enhance BD production and consumption. The plan includes:

1. Revival of tax breaks in 2012,
2. Mandatory blending of BD with commercial diesel,
3. Collection system for waste oils from home and food industry, and
4. Support for BD R&D.

An interesting investment by the Korean government was observed in active involvement in Jatropha farm development in foreign countries. In 2009, a Korean company developed 1.1 million hectare of Jatropha farmland, and currently another 1.37 million hectare farm is being developed.

8.3 Direction of Technology

The significant factors that affect the cost of BD are feedstock cost, plant size, and value of the glycerine by-product (Kulkarni and Dalai, 2006) [9]. One of the most crucial variables that affect the cost of BD is the cost of the raw materials. Waste cooking oil (WCO), which is much less expensive than pure vegetable oil, is a promising alternative to vegetable oil for BD production. Restaurant waste oils and rendered animal fats are less expensive than food-grade canola and soybean oil. Currently, all these waste oils are sold commercially as animal feed. However, since 2002, the European Union (EU) enforced a ban on feeding these mixtures to animals, because, during frying, many harmful compounds are formed and, if the WCO is used as an additive to feeding mixtures for domestic animals, then it could result in the return of harmful compounds back into the food chain through the animal meat. Hence, the WCO must be disposed safely or be used in a way that is not harmful to human beings. The quantity of WCO generated per year by any country is huge. The disposal of WCO is problematic, because disposal methods may contaminate environments including drinking water sources. Many developed countries have set policies that penalize the disposal of waste oil through the water drainage.

The production of BD from WCO is one of the better ways to utilize it efficiently and economically. The data on the requirements of diesel fuel and availability of WCO in any country indicate that the BD obtained from WCO may not replace diesel fuel completely. However, a substantial amount of diesel fuel can be prepared from WCO, which would partly decrease the dependency on petroleum-based fuel. The amount of WCO generated in each country varies, depending on the use of vegetable oil. An estimate of the potential amount of WCO collected in the EU is 700,000–1,000,000 tonnes/yr. On an average, 9 pounds of yellow grease per person were produced annually in the US. 120,000 tonnes/yr of yellow grease is produced in Canada.

Note that the WCO/waste fryer grease (WFG) is categorized by its FFA content. For example, if the FFA content of WCO is <15%, then it is called "yellow grease"; otherwise, it is called "brown grease". Because of the production of such large quantities of WCO, a substantial amount of BD can be produced from this material.

A preliminary case study conducted on the requirement of BD in Canada to meet the B5 requirement reported that 610 million liters of the BD are required per year. Hence, a substantial portion of the BD (of the 5% requirement in Canada) can be replaced by the BD obtained from WCO. The properties of the BD from WCO would be largely dependent on the physicochemical properties of these forms of feedstock.

8.4 Case Study: Strategy for Korea

It is noted that Korea is rapidly moving to expand its current BD infrastructure. The future R&D efforts should be directed towards improving technologies for municipal solid waste (MSW), farm waste, and algae.

Current efforts to produce feedstock are well-planned and will be helpful. MSW and farm waste will help in reducing the cost of BD. This strategy will work regardless of the price fluctuation of BD in the market.

8.5 Case Study: Improvement of Current Policy of Korea

Our analysis indicates that Korea has taken the right steps in order to build BD infrastructure. The policy in the coming years should encourage the use of MSW and farm waste as a feedstock to produce BD. Production of BD from Korean grown algae should be explored by developing new technologies.

In summary, the following policies will increase production of BD in Korea:

- Continue tax incentives for developing BD infrastructure.
- Encourage the use of non-edible crops for BD production.
- Encourage the use of municipal solid waste and farm waste to produce BD.
- Encouraging research incentives through grants and tax rebates for developing new and improved technologies for the use of algae, municipal solid waste, and farm waste.
- Study the environmental impact of BD produced from different feedstock.

8.6 Case Study: Impacts of Policy in South America

a. Argentina

In Argentina, diesel is required to contain a 5% biodiesel share beginning in January 2010. This is in accordance with Argentine Biofuel Law 26.093 of April 2006, which was implemented under Decree 109/2007 in February 2007. Biodiesel's quality specification was formalized with Resolution 6/2010 by February 2010 and its domestic price was set up with Resolution 7/2010 in February 2010. These laws owe to a recent increase in biodiesel output and productive capacity (an output increment of 433% taking place in 2008, as reported by the Argentine Renewable Energies Chamber, 2009).

The country's main oil-bearing crop is soybean, which accounts for 80% of the oil-bearing plant production by volume. In April 2006, "Senator Falco's Law" was introduced in the country, which provided a basis of mandatory blending of 5% of biodiesel into diesel. It was projected that the demand of biodiesel would grow annually in 2010–2020. This demand was estimated to be 700 million liters in 2010 and it is projected to be around 987 million liters in 2020 (Avinash et al., 2014) [10]. The government stimulated the domestic market of biodiesel in 2006. Law No. 26093 established the regulation and promotion of sustainable biodiesel use through the mandatory addition of biodiesel to fossil diesel. In 2010, the blend was already of 5% of biodiesel in petroleum diesel.

The introduction of the biodiesel blend along with the large boost in Argentine biodiesel exports had decreased greenhouse gas emissions by over 4 million tons in 2011 (Timilsina & Shrestha, 2011) [11]. In addition, Forestry Law No. 26.331 was passed in 2007, which in theory should control indirect deforestation from the expansion of soy farming and provide payment for some ecosystem services, but thus far implementation has been weak (Law No. 26.331 of Argentina, 2007). The onset of Argentina's biodiesel policy dates back to Resolution 129 in 2001, where biodiesel was defined. Under the Biofuel Law 26093, manufacturers of biodiesel have three options:

- To produce for self-consumption or the domestic market.
- To take advantage of several tax and investment incentives.
- To produce for export and thus not be eligible to receive incentives.

To receive any incentives available for biodiesel, companies must hold operations in Argentina and only be dedicated to biodiesel production, with the company's equity mainly handled by agricultural producers or the government. Biodiesel governed by this promotional regime are subsidized by exemption from fossil fuel taxes. Regulations for implementing Law 26.093 were issued in 2007 by Decree 109. Incentives for production and use of biodiesel in the domestic market were created with a 15-year timeframe, starting in 2010.

Argentina's biodiesel policy development has been highly active since 2010, with several resolutions of the Secretariat of Energy published in the Official Gazette. The target quality features that biodiesel should meet were specified and subsequently modified by Resolutions 6 and 828, respectively. Ratification of the Biodiesel Supply Agreement and establishment of a polynomial formula to calculate the price of biodiesel were addressed in Resolution 7.

The Secretariat of Energy issued Resolution 554 in 2010, which increased the mandatory use of biodiesel with diesel from 5% to 7%. This was increased in 2013 to 8%, and a target of 10% was set for 2014, at the end of 2013. Initially, biodiesel export taxes and refunds were low. Biodiesel export taxes started at 5%, and increased to 20% by 2008. This rate was less than half the rate applied to soybean oil, and also allowed a 2.5% refund.

Lower taxes boosted large investment in the biodiesel industry, which ended up with production of 2.45 million tons of biodiesel, 1.52 million tons of which were exported in 2012 (Demirbas, 2009) [12]. Conversely, biodiesel export taxes were increased and refunds eliminated, as stated in Decree 1339 of August 2012, which raised the biodiesel export tax rate to 32%. Another problem was that the biodiesel benchmark price may have been set too low by the government in 2012. An Executive Interdisciplinary Monitoring Unit calculated the reference price paid in the internal market. First, Resolution 1436 issued by the Secretariat of Energy set the price at $4.405/ton. Next, Resolution 1725 required companies that blended biodiesel with petroleum to pay $4.661/ton of biodiesel to small biodiesel processing companies. Because the cost to produce one ton of biodiesel exceeded $5, Resolution 1725 extended the crisis that had begun with enactment of Resolution 1436. Given these circumstances, the government also segmented the domestic prices of biodiesel, establishing different values for small, medium, and large companies by December 2012. This scheme allowed the small biodiesel processing companies to recover the same productivity levels they had before the crisis.

There has been no non-refundable economic contribution from the National Treasury to subsidize biodiesel. Moreover, the domestic mandate of a 10% blend diverts part of the biodiesel output and feedstock away from the higher economic returns possible in the EU market. The results of the biodiesel mandates in Argentina are near to or exceed the policy goals. Biodiesel use was at 7.3% in 2012, though it fell in 2013 below the 8% mandate because of the conflict with the EU. While small biodiesel processing companies work at full capacity due to the paralysis of the biodiesel export industry, the total industry used just 35% of its installed production capacity in 2013. As a result, biodiesel output fell significantly in 2013 along with export earnings, though in response the domestic mandate was increased to 10% for 2014 (Solomon et al., 2015) [13]. Taken together, these circumstances highlight the underlying role that energy policy makers play in the economy and international trade of a country. In 2017, Argentina produced the fourth largest amount of biodiesel, 3.3 billion liters per year, whereas the US produced 6 billion liters per year, Brazil 4.3 billion liters per year, and Germany 3.5 billion liters per year.

b. Brazil

The government of Brazil inaugurated the National Biodiesel Production and Use Program (PNPB) in 2005 to develop a biodiesel sector for the country. Federal law 11.097/05 of 2003 officially started the biodiesel program, primarily based on soy oil. This was carried out partly because of agreements between the government and manufacturers to develop a market of purposely-modified vehicles. Initially, PNPB mandated that 2% of petrol-based diesel be replaced by biodiesel by 2008. These blending mandates were further increased and set to 5% in 2010 (Hieltjes and Lahiri, 2009) [14]. By introducing a biodiesel sector that is subsidized and not cost-competitive, the government hoped to further reduce dependence on foreign oil and to cut greenhouse gas emissions. The output capacity reaches a total of 4 billion liters per year. Although soybean is the primary base of biodiesel production, other vegetable oil plants are used: palm tree, castor bean, and Jatropha.

The Brazilian government aims to make the production of biodiesel a tool for social inclusion in family farming, with the developing and disseminating of crops adapted to the conditions of each region of the country. This program had the goal to reach one billion liters a year, starting from 2008. At one point, the country had 55 companies for biodiesel production and 65 other associated companies involved in the program, which provided 30,000 rural jobs (Da Costa et al., 2010) [15]. The southern region also obtains feedstock from family farming through PNPB, since a significant number of establishments of this kind of agriculture are organized in cooperatives. In 2011 Brazil stood out in biodiesel production in the world, representing 11.41% of the global production.

Another goal of PNPB was to reduce regional socioeconomic disparities by enabling family soybean farmers to participate in the bio-economy. Soy oil accounts for around 73% of the biodiesel feedstock, with the rest coming from beef fat, tallow, and other fats and oils (Canoira et al., 2008) [16]. The goals of the PNPB program to develop biodiesel in Brazil were, therefore, based on concerns for advancing social, economic, and environmental sustainability of the domestic interest in biodiesel. In addition, in 2004 the Selo Social (Social Stamp) program was created by Decree No. 5297 to identify family farmers who could participate in the bio-economy and thereby increase employment opportunities. In order to encourage the infant biodiesel industry in the country, the Brazilian government has provided tax exemption for biodiesel depending on the type of raw material used, region of production, and size of farms growing raw materials. The distilleries and bio-refineries that are involved in buying raw materials under the Selo Seal program are better suited to get credit lines from national banks to obtain the right to sell biodiesel to the National Petroleum agency, and also receive other benefits. In 2017, Brazil achieved the second largest biodiesel production in the world producing 4.3 billion liters per year next to the US that produced 6 billion liters per year (Naylor and Higgins, 2017) [17]. Table 8.2 provides information about the possible biodiesel feedstocks used in different South American countries.

c. Chile

The focal point on first-generation biodiesel in Chile is not promising, so the country has to look forward to producing biodiesel from second-generation biodiesel crops like residual wood from exotic species of Radiata pine and eucalyptus. Also, the new research and development projects have to be

TABLE 8.2

Possible biodiesel feedstocks in South American countries

#	Nation	Feedstock
1.	Argentina	Soybean, Sunflower, Crambe Abyssinica, Jatropha Macrocarpa, Cuphea
2.	Brazil	Soybean, Palm, Canola, Castor, Pachira glabra, Jatropha curcas L
3.	Chile	Residual wood (from radiata pine and Eucalyptus), Camelina, Rapeseed
4.	Colombia	Palm, Acrocomia aculeate
5.	Paraguay	Palm, Acrocomia aculeate, Tung, Jatropha curcas L
6.	Peru	Palm, Jatropha
7.	Uruguay	Coconut, Beef tallow, rice and vegetable oil
8.	Venezuela	Palm, Coconut, Acrocomia aculeate

proposed on biomass conversion technologies in Chile. In contrast, low-cost prospective feedstock like Camelina could be cultivated in Chile.

Biodiesel integrated the class of fuels through Law No. 20339/2009. The use of biodiesel was supported by Decree No. 11/2008, which authorizes the blend of 2–5% of biodiesel with diesel, but the blend is facultative. Moreover, Circular No. 30 exempts biodiesel from taxes. In 2009, the capacity for biodiesel production in Chile was about 75,000 m^3 (75 million liters). The capacity in the three previous years was less than 3,000 m^3 (3 million liters) (Cabal et al., 2011) [18]. Thus, a rapid expansion in the biodiesel production sector can be noticed. Even so, the current production represents only 37% of the minimum necessary for the implementation of the policy for an additional 2% of biodiesel to fossil diesel.

d. Colombia

In Colombia, the biodiesel produced is mostly intended for the transportation sector, aiming to gradually replace fossil fuels used in the vehicles operating in the country. In 2008, the blend of 5% of biodiesel with petroleum diesel was introduced in the domestic market (Sorda et al., 2010) [19]. The Colombian biodiesel production is chiefly based on palm oil, as well as other crops such as cotton, soybean, sesame, and other oil seeds not intended for human consumption like castor bean and Jatropha curcas. Palm oil has great relevance for the Colombian industry, and it has represented 8% of the total worldwide production per year in the period of 2008–2010 (Furumo and Aide, 2017) [20]. Colombia occupies the fifth place in palm oil production in the world and the first place in Latin America. The biodiesel is added to diesel at 7–10%, depending on the region of the country. Biodiesel is transported in oil pipelines mixed with diesel in proportions lower than 4%. Table 8.3 provides most recent statistics of biodiesel for South America.

TABLE 8.3

Biodiesel Statistics for South America

	Production (million L)		Growth (%) (1)	Domestic use (million L)		Growth (%) (1)	Share in Diesel use (%)				Net Trade (million L) (2)	
							Energy share		Volume share			
	Avg 2012-14est	2024	2015–24	Avg 2012-14est	2024	2015–24	Avg 2012-14est	2024	Avg 2012-14est	2024	Avg 2012-14est	2024
Argentina	2565	2923	1.17	1043	1429	0.62	6.7	9.5	7.3	10.3	1522	1494
Brazil	3118	5094	1.23	3119	5070	1.19	4.9	6.5	5.3	7.0	-1	24
Colombia	666	968	3.34	665	968	3.37	..	..	..	..	1	-0
Peru	98	108	0.03	275	272	1.57	..	..	..	..	-177	-165

Note: Average 2012-14est: Data for 2014 are estimated.

1. Least-squares growth rate.
2. For total net trade, sum of all positive net trade positions.

Source: OECD/FAO (2015) [21], "OECD-FAO Agricultural Outlook," OECD Agriculture statistics (database). doi: dx.doi.org/10.1787/agr-outl-data-en

The promotion of biodiesel production and consumption of biodiesel began with both Law 939 of 2004 and Resolution 1289 of 2005. The latter instituted a requirement for 5% biodiesel blend by 2008. Future plans include increasing the blend to 10% in 2010 and 20% in 2012. Policies are in place to support the local biodiesel industry by providing incentives to agro-industrial projects. Meanwhile, concerns over the impact of biodiesel on food prices and greenhouse gas emissions have taken place within political debate, with new rules guiding the support given to biodiesel.

e. Paraguay

In 2007, Paraguay's Ministry of Industry and Trade determined that 1% of biodiesel should be added to fossil diesel. These values increased to 3% and 5% in 2008 and 2009, respectively. With the consumption of diesel in the country, 50 million liters of the biodiesel would be needed to meet the demand of the national fleet in proportions of addition of 5% of biodiesel to conventional diesel (Cremonez et al., 2015) [22]. Among the main energy crops liable to cultivation and production of biodiesel in Paraguay, there are castor bean, soybean, sunflower, tung tree, peanut, cotton, Jatropha curcas, canola, animal fat, and residual oils.

f. Peru

In Peru, palm oil and Jatropha curcas are the two major feedstocks that provide an opening to develop biodiesel market at the national level. The excellent agroclimatic conditions of Peruvian Amazonian jungles offer a basis for the planting of palm trees and Jatropha. In 2003, Peru adopted policies to stimulate the national biodiesel market, seeking the development of alternate energy sources. Biodiesel started to be added in a proportion of 5% to diesel in 2011.

g. Uruguay

Uruguay is among the countries being able to produce fuels from energy crops. In 2002, Law No. 17567 promoted the production of alternate fuels that could substitute fossil fuels using local materials of plant and animal origin. The National Institute of Agricultural Technology (INTA) developed a project for biodiesel production from different vegetable oils, primarily based on the production of rapeseed oil. This initiative had Brazil as a role model, which sought to develop raw materials for biodiesel production that were suitable for each region of the country in economic, environmental and social aspects.

In addition to the state-run research, many private ventures have been in operation for biodiesel production, as well as use of vegetable oil and beef tallow. However, the installed capacity of production in these enterprises is low and most of the investments are low in scale and come from foreign companies. Furthermore, the economic viability of biodiesel is directly related to the price of inputs and final value of the feedstock. This takes into account that the serious economic problems for price fluctuations of these components could cause serious economic problems for these companies.

h. Venezuela

Venezuela's biodiesel production is considered as en route to the second generation biodiesel. Biodiesel from first generation routes are normally produced from the cultivation of food crops and could create competition of the production of biodiesel with the production of food and use of water in the country. Fuels of second and third generations are produced from ligno-cellulosic residues from forests, agro-industries, and grasses of short rotation. Although the prospects are not as positive as for the other countries in South America, the overall performance of Venezuela shows tendencies to increase from both economic and environmental perspectives.

8.7 Concluding Remarks

According to the BP statistical review for world energy (2017), global biofuels production rose by 2.6% in 2016, well below the 10-year average of 14.1%, but faster than in 2015 (0.4%). Biodiesel production rose by 6.5% with Indonesia providing more than half of the increment (1,149 thousand tonnes of oil equivalent or ktoe). International Energy Outlook in 2017 reported that production of liquid fuels, including biofuels, increases by 25% from 2015 to 2040. The use of refined petroleum and liquid fuels in the transportation sector will continue to increase through 2040. However, their share decreases from 95% to approximately 88% as the use of alternative fuels gradually increases.

As presented in this book, the two major BD-producing countries, Germany and the United States, have been cutting the government subsidies and tax breaks for BD manufactures and users since 2008. Other large BD producing countries such as France and Italy are also following the same steps. Currently without these assistances, BD is not competitive enough to increase the market share over gasoline or regular diesel. Current economic crisis, which brought about termination of BD subsidies, exacerbates BD business. Therefore, BD is under huge pressure coming from two different problems that has the identical cause.

Even under these circumstances, BD production and consumption are bouncing back. This is a greatly promising sign that previews BD potential in the future. When BD price becomes more competitive due to broadening of feedstock sources and advancement of production technology, BD is believed to excel itself tremendously.

In many research and studies, it has been strongly asserted that renewable fuels production is a source of job creation, economic activity and higher revenues. For example, a sweeping and independent study conducted by econometric firm Doyletech Corporation in Canada, concluded that the renewable fuels sector in Canada has provided a $2.9 billion boost to economic activity and generated some 14,000 full-time jobs. What's more, each and every year it is responsible for further $1.5 billion in economic activity and an additional 1,000 new jobs. The report studied 28 ethanol and BD plants across Canada and added that there were major benefits from renewable fuels in "rural re-vitalization, increased oil exports from western Canada, industrial development, and valuable options for re-balancing fuel 'mix'." Furthermore, the US biodiesel industry supports nearly 64,000 jobs across the US with more than $2.5 billion in wages, and an overall economic impact of nearly $11.5 billion. It is produced all across the country in all regions – achieving a level of success for this industry.

Various feedstock sources are being developed and used depending on the regional characteristics and economic activities. Plant oils such as rapeseed, canola, Jatropha, soybean and palm oil are proven to efficient and have no or a very little negative impact on food. Waste oils should be used more and appear to be increasing in BD manufacturing. Algae is another promising feedstock source for BD. Feedstock types, their impacts, and related BD production technologies are addressed in more detail in Chapters 4 (Biodiesel production technologies), 5 (Algae biodiesel), and 6 (Properties of biodiesel from different feedstock types).

Some aspects of oil company's activities in BD area are reviewed in this chapter. It is encouraging to see major oil companies are investing in BD. It is thought that company involvement will boost BD production, will help lower the price, and will help advance the technology. As a result, it will help BD have greater portion in energy production and consumption.

In summary, the biodiesel policies are significantly influenced by energy production, energy needs, availability of feedstock, environmental issues, and tax structure of a nation, which in turn are influenced by the domestic and diplomatic political situations, and economic status of the country.

References

1. IEA Bioenergy. (2007). Potential contribution of bioenergy to the world's future energy demand. IEA Bioenergy: ExCo 2007 02–12.
2. Digman B., Joo H. S., & Kim D.-S. (2009). Recent progress in gasification/pyrolysis technologies for biomass conversion to energy. Environmental Progress & Sustainable Energy, 28(1), 47–51.
3. Ail, S. S., & Dasappa, S. (2016). Biomass to liquid transportation fuel via Fischer Tropsch synthesis–Technology review and current scenario. Renewable and Sustainable Energy reviews, 58, 267–286.

4. Doornbosch, R., & Steenblik, R. 2007. OECD round table on sustainable development: Biofuels: Is the cure worse than the disease? September.
5. http://petroleum.nic.im/biodiesel.pdf
6. www.ops.gov.ph/records/ao_no103.htm
7. Lim, S., & Teong, L. K. (2010). Recent trends, opportunities and challenges of biodiesel in Malaysia: An overview. Renewable and Sustainable Energy Reviews, 14, 938–954.
8. Bart, J. C. J., Palmeri, N., & Cavallaro, S. (2010). Biodiesel science and technology from soil to oil. Woodhead publishing series in energy (first ed.), Vol. 7, CRC Press, Boca Raton, Boston, New York, Washington, DC.
9. Kulkarni, M. G., & Dalai, A. K. (2006). Waste cooking oils an economical source for biodiesel: A Review. Ind. Eng. Chem. Res., 45, 2901–2913.
10. Avinash, A., Subramaniam, D., & Murugesan, A. (2014). Bio-diesel—A global scenario. Renewable and Sustainable Energy Reviews, 29, 517–527.
11. Timilsina, G. R., & Shrestha, A. (2011). How much hope should we have for biofuels? Energy, 36(4), 2055–2069.
12. Demirbas, A. (2009). Political, economic and environmental impacts of biofuels: A review. Applied Energy, 86, S108–S117.
13. Solomon, B. D., Banerjee, A., Acevedo, A., Halvorsen, K. E., & Eastmond, A. (2015). Policies for the sustainable development of biofuels in the Pan American region: A review and synthesis of five countries. Environmental Management, 56(6), 1276–1294.
14. Hieltjes, T., & Lahiri, S. (2009). Explorative research on the transport biofuel market.
15. Da Costa, A. C. A., Junior, N. P., & Aranda, D. A. G. (2010). The situation of biofuels in Brazil: New generation technologies. Renewable and Sustainable Energy Reviews, 14(9), 3041–3049.
16. Canoira, L., Rodríguez-Gamero, M., Querol, E., Alcántara, R., Lapuerta, M., & Oliva, F. (2008). Biodiesel from low-grade animal fat: Production process assessment and biodiesel properties characterization. Industrial & Engineering Chemistry Research, 47(21), 7997–8004.
17. Naylor, R. L., & Higgins, M. M. (2017). The political economy of biodiesel in an era of low oil prices. Renewable and Sustainable Energy Reviews, 77, 695–705.
18. Cabal, H., Labriet, M., & Lechón, Y. (2011). Chapter 12: Review of the world and European renewable energy resource potentials. Handbook of Sustainable Energy, 244–269.
19. Sorda, G., Banse, M., & Kemfert, C. (2010). An overview of biofuel policies across the world. Energy policy, 38(11), 6977–6988.
20. Furumo, P. R., & Aide, T. M. (2017). Characterizing commercial oil palm expansion in Latin America: Land use change and trade. Environmental Research Letters, 12(2), 024008.
21. OECD/FAO. (2015). OECD-FAO agricultural outlook. OECD Agriculture statistics (database). doi: dx.doi.org/10.1787/agr-outl-data-en.
22. Cremonez, P. A., Feroldi, M., Feiden, A., Teleken, J. G., Gris, D. J., Dieter, J., de Rossi, E., & Antonelli, J. Current scenario and prospects of use of liquid biofuels in South America. *Renewable and Sustainable Energy Reviews*, 43 (2015), 352–362.

Index